Sea Cucumbers

Sea Cucumbers
Aquaculture, Biology and Ecology

Mohamed Mohsen

Institute of Oceanology,
Chinese Academy of Sciences,
Qingdao, P.R. China
Faculty of Agriculture, Al-Azhar University,
Nasr City, Cairo, Egypt

Hongsheng Yang

Institute of Oceanology,
Chinese Academy of Sciences,
Qingdao, P.R. China
University of Chinese Academy of Sciences,
Beijing, P.R. China

ELSEVIER

ACADEMIC PRESS
An imprint of Elsevier

Academic Press is an imprint of Elsevier
125 London Wall, London EC2Y 5AS, United Kingdom
525 B Street, Suite 1650, San Diego, CA 92101, United States
50 Hampshire Street, 5th Floor, Cambridge, MA 02139, United States
The Boulevard, Langford Lane, Kidlington, Oxford OX5 1GB, United Kingdom

Notices
Knowledge and best practice in this field are constantly changing. As new research and experience broaden our understanding, changes in research methods, professional practices, or medical treatment may become necessary.

Practitioners and researchers must always rely on their own experience and knowledge in evaluating and using any information, methods, compounds, or experiments described herein. In using such information or methods they should be mindful of their own safety and the safety of others, including parties for whom they have a professional responsibility.

To the fullest extent of the law, neither the Publisher nor the authors, contributors, or editors, assume any liability for any injury and/or damage to persons or property as a matter of products liability, negligence or otherwise, or from any use or operation of any methods, products, instructions, or ideas contained in the material herein.

Library of Congress Cataloging-in-Publication Data
A catalog record for this book is available from the Library of Congress

British Library Cataloguing-in-Publication Data
A catalogue record for this book is available from the British Library

ISBN: 978-0-12-824377-0

For information on all Academic Press publications visit our website at
https://www.elsevier.com/books-and-journals

Publisher: Charlotte Cockle
Acquisitions Editor: Patricia Osborn
Editorial Project Manager: Allison Hill
Production Project Manager: Sreejith Viswanathan
Cover Designer: Alan Studholme

Typeset by TNQ Technologies

Contents

Preface ix
Acknowledgements xi

CHAPTER 1 Introduction to sea cucumbers 1
   1.1 Definition 1
   1.2 Body of sea cucumbers 1
   1.3 Tube feet and papillae 3
   1.4 Mouth and anus 4
   1.5 Extraordinary animals 6
   1.6 Treasure of the seabed 8
   References 13

CHAPTER 2 Anatomic structure and function 19
   2.1 Body walls 19
   2.2 Ossicles and calcareous ring 19
   2.3 Circulatory system 23
      2.3.1 Haemal system 23
      2.3.2 Coelomic fluid and coelomocytes 24
   2.4 Digestive system 24
      2.4.1 Morphology 24
      2.4.2 Histology 26
   2.5 Respiratory system 26
      2.5.1 Morphology 26
      2.5.2 Histology 27
   2.6 Water vascular system 28
   2.7 Nervous system 29
   2.8 Reproductive system 29
   References 34

CHAPTER 3 Behaviour and ecology 37
   3.1 Evisceration 37
   3.2 Regeneration 38
   3.3 Aestivation 42
   3.4 Population genetics 43
   3.5 Sea cucumbers interaction with the surrounding environment ... 47
      3.5.1 Bioturbation 47
      3.5.2 Benthic oxygen regeneration 48
      3.5.3 Organic matter decomposition 48
      3.5.4 Algal bloom cleaning 48

3.5.5 Water chemistry improving 49

3.5.6 Mariculture waste bioremediation 49

3.5.7 Relationship with benthic organisms 50

3.5.8 Predators 53

References 54

CHAPTER 4 Sea cucumbers research in the Mediterranean and the Red Seas 61

4.1 Region under study 61

4.2 Biology and ecology of sea cucumber species 63

4.2.1 Sea cucumbers in the Mediterranean Sea 63

4.2.2 Sea cucumbers in the Red Sea 69

4.3 Sea cucumbers aquaculture development 83

4.3.1 Asexual reproduction 83

4.3.2 Sexual reproduction 84

4.4 Sea cucumber utilisation 85

4.4.1 Ecological values 85

4.4.2 Nutritional and medicinal values 87

4.5 Conclusion remarks 93

References 94

CHAPTER 5 Sea cucumbers research in the Persian Gulf 103

5.1 Region under study 103

5.2 Biology and ecology of sea cucumber species 104

5.2.1 *Stichopus hermanni* 104

5.2.2 *Holothuria leucospilota* 106

5.2.3 *Holothuria scabra* 106

5.2.4 *Holothuria hilla* 107

5.2.5 *Holothuria impatiens* 108

5.2.6 Other species 108

5.3 Aquaculture development 110

5.4 Sea cucumbers utilisation 111

5.4.1 Nutritional and medicinal values 111

5.5 Conclusion 120

References 120

CHAPTER 6 Sea cucumbers mariculture 127

6.1 Broodstock collection 127

6.2 Gonadal development 128

6.3 Broodstock conditioning 130

6.4 Artificial induction of spawning 131
    6.4.1 Thermal shock 131
    6.4.2 Gonadal stimulation 132
    6.4.3 Drying 132
    6.4.4 Mechanical shock 132
    6.4.5 Water pressure plus temperature 132
    6.4.6 Algae path 132
6.5 Spawning behaviour 133
6.6 Fertilisation 135
6.7 Embryo, larval and juvenile development 137
6.8 Environmental factors 143
6.9 Grow-out methods 144
    6.9.1 Earthen ponds 145
    6.9.2 Sea pens 145
    6.9.3 Suspended culture 145
    6.9.4 Co-culture 146
6.10 Feasibility for sea cucumbers farming 149
References 151

CHAPTER 7 Sea cucumbers processing and cooking 157
7.1 Processing sea cucumbers 157
    7.1.1 Drying 157
    7.1.2 Instant sea cucumbers 162
    7.1.3 Canned sea cucumbers 164
7.2 Cooking 164
    7.2.1 Braised sea cucumbers 165
    7.2.2 Sea cucumber salad 166
    7.2.3 Sea cucumber with mushrooms 166
    7.2.4 Deep-fried sea cucumber 166
    7.2.5 Braised sea cucumbers with vegetables 167
    7.2.6 Braised sea cucumber with scallions 168
References 169

CHAPTER 8 Developing sea cucumbers aquaculture in the
                  Middle East: a perspective 173
8.1 Background 173
8.2 Potential species for aquaculture 175
8.3 Potential areas for sea cucumber mariculture 176

8.4 Potential models for sea cucumber mariculture.................... 177

8.5 Sea cucumber marketing.. 179

8.6 Research plan 180

References... 181

Index.. 185

Preface

Sea cucumbers have gained considerable attention because of their beneficial influence on human health and possible therapy. In addition to their nutritional and medicinal values, sea cucumbers play an important role in marine ecosystems. Sea cucumbers' activities improve sediment characteristics, enhance water chemistry, remediate aquaculture wastes and influence the productivity of many benthic organisms. With the increase of the market demand, overharvesting sea cucumbers has been extended from Asia to the world. Given the harvesting ease and slow population regeneration of sea cucumbers, many sea cucumber species have been extirpated. Therefore, there is an urgent need for sea cucumber aquaculture and restocking programmes to cover the market demand and prevent the extinction of species. To meet the challenge, there is a need to share experience across regions and facilitate learning so that mistakes are not repeated. *Sea cucumbers: Aquaculture, Biology and Ecology* is a reference book that gathers practical and biological knowledge necessary to promote aquaculture of sea cucumbers. Studying biology, reproduction conditions and optimum grow-out strategies is essential for the development of a successful aquaculture product. The book pays particular attention to the research of sea cucumbers in the Middle East, where experience with sea cucumber is limited but of potential value.

The book starts with the biology of sea cucumbers, presenting the fundamental biological and ecological aspects of sea cucumbers as well as their nutritional and medicinal use and value. The second part of this book aims to shed light on the lesser mentioned resources of sea cucumber in the Middle East. We summarised the available knowledge on sea cucumbers in the Mediterranean Sea, the Red Sea and the Persian Gulf, particularly the variety of sea cucumber species, aquaculture development and utilisation research of sea cucumbers. The third part of this book aims to summarise the available knowledge related to the aquaculture of six holothurians that exist in the Middle East, presenting the knowledge needed to promote their aquaculture and fisheries management. Also, the last part includes processing and cooking methods of sea cucumbers for an effective marketing process. Finally, we discuss what we need to know to improve sea cucumbers aquaculture in the Middle East.

After Mohsen started his PhD in China, it took about 36 months of collecting, writing and revising with internal reviewers until the book was finally ready. We received comments from five experts in sea cucumbers research who devoted their time and efforts to improve the quality of this book. They are Dr Xiutang Yuan (Yantai Institute of Coastal Zone Research, Chinese Academy of Sciences), Dr Muyan Chen (Ocean University of China), Dr Zonghe Yu (South China Agriculture University), Dr Tianming Wang (Zhejiang Ocean University) and Dr Peng Zhao (Hainan

University). We desire that this book will stimulate further research and development on the sea cucumber aquaculture and stock enhancement in the Middle East or elsewhere. Success in this endeavour is paramount to the very survival of the marine ecosystems.

Dr Mohamed Mohsen
Institute of Oceanology,
Chinese Academy of Sciences,
Qingdao, P.R. China
Faculty of Agriculture, Al-Azhar University,
Nasr City, Cairo, Egypt

Dr Hongsheng Yang
Institute of Oceanology,
Chinese Academy of Sciences,
Qingdao, P.R. China
University of Chinese Academy of Sciences,
Beijing, P.R. China

Acknowledgements

We would like to thank the researchers who shared their images to support the content of this book: Dr Mohamed Ismael, Dr Mehmet Aydin and Dr Aymeric Desurmont together with many passionate photographers who made their work accessible online. Also, we would like to thank researchers on sea cucumbers around the world for the many publications that made this book possible.

Some special thanks go to our lab members (Institute of Oceanology) for advising on the content of this book and support: Dr Yi Zhou, Dr Tao Zhang, Dr Libin Zhang, Dr Lina Sun, Dr Shilin Liu, Dr Chenggang Lin, Jingchun Sun, Dr Lili Xing, Beibei An, Dr Feng Jie, Dr Da Huo, Dr Hao Song, Dr Xiaoshang Ru, Dr Xiaoyue Song, Dr Ding Kui, Xin Xiaoke, Dr Li Xiaoni, Yang Meijie and Shidong Yue.

Finally, we would like to thank Elsevier team for their assistance in producing this book. Also, Mohamed Mohsen would like to thank his parents, family members and his friends for their generous support.

Introduction to sea cucumbers

1.1 Definition

Sea cucumbers, also known as holothurians, are from the phylum Echinodermata. From ancient Greek, echinos means a hedgehog and derma means skin. Sea cucumbers are from the class Holothuroidea with a leathery muscular and elongated body, with tentacles surrounding the anterior end. They can be found in all ocean regions, from the polar to tropical and from the deep ocean to intertidal (Conand, 2006). Sea cucumbers are classified depending on the shape and the existence of the tentacles, respiratory tree, ossicles, papillae, tube feet and, recently, using molecular phylogeny; however, understanding the overall systematics of the group remains uncertain (Smirnov, 2012; Miller et al., 2017). The widely used classification of sea cucumbers is defined by Pawson and Fell (1965) according to the tentacles, body morphology and, partially, the shape of the calcareous ring (Pawson and Fell, 1965). Six orders are classified based on that: Molpadiida, Apodida, Aspidochirotida, Elasipodida, Dendrochirotida and Dactylochirotida. There are about 1717 identified species as holothurians worldwide (Paulay and Hansson, 2013). Among them, a total of 58 sea cucumber species are considered commercially important (Purcell et al., 2012). Most commercial sea cucumbers belong to the order Aspidochirotida, and few belong to the order Dendrochirotida (Conand, 2006).

1.2 Body of sea cucumbers

The body of sea cucumbers is longitudinally symmetric, cylindric and elongated primarily with leathery skin. Their colours are brown, grey, purple, white, red, orange or violet (Figs. 1.1–1.5) (Burton and Burton, 2002). Commonly, their size ranges from 10 to 30 cm in length, and some species can reach 5 m long (Piper, 2007). The thickness of the body wall indicates the commercial importance of the species because it is processed for human consumption (Purcell et al., 2012).

FIGURE 1.1

The curry fish sea cucumber *Stichopus herrmanni* seen in Mayotte lagoon.

FIGURE 1.2

The sea cucumber *Stichopus chloronotus* seen in Philippine Islands, Occidental Mindoro, Apo Reef.

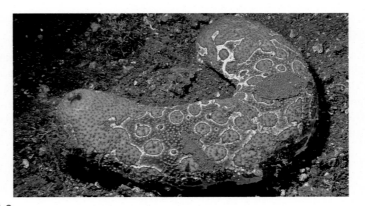

FIGURE 1.3

The sea cucumber *Bohadschia ocellate*.

FIGURE 1.4

The sea cucumber *Colochirus robustus.*

Credit: Nhobgood Nick Hobgood. Licensed under CC BY-SA 3.0 via Wikimedia Commons.

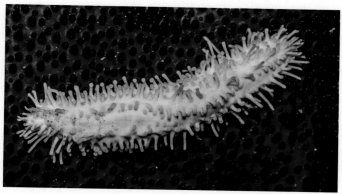

FIGURE 1.5

The sea cucumber *Eupentacta quinquesemita.*

Credit: Jerry Kirkhart. Licensed under CC BY 2.0 via Wikimedia Commons.

1.3 Tube feet and papillae

The ventral surface has tube feet, while the dorsal surface has papillae (Woo et al., 2015) (Fig. 1.4). The tube feet are cylindric with a sucker at the top (Fig. 1.6); they are used for locomotion and adhesion. The papillae provide a protection function when sea cucumbers stiffen their body wall (Motokawa et al., 1985). The localisation and numbers of the tube feet and papillae vary among sea cucumber species (Díaz-Balzac et al., 2010). The papillae are unnoticeable in the orders Dendrochirotida, Apodida and Molpadiida, whereas the papillae can be detected in the orders Elasipodida and Aspidochirotida (Chang et al., 2011). The tube feet may be scattered evenly on the ventral surface of the body or distributed in five lateral grooves (Woo et al., 2015).

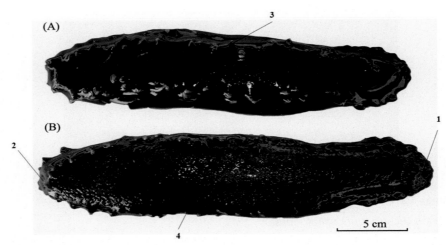

FIGURE 1.6

Dorsal (A) and ventral (B) views of the sea cucumber *Stichopus chloronotus*. 1, the mouth and the tentacles (retracted); 2, the anus; 3, the papillae; 4, the tube feet.

Modified from Woo, S.P., Yasin, Z., Tan, S.H., Kajihara, H., Fujita, T., 2015. Sea cucumbers of the genus Stichopus Brandt, 1835 (Holothuroidea, Stichopodidae) in Straits of malacca with description of a new species. ZooKeys (545), 1. Licensed under CC-BY 4.0.

1.4 Mouth and anus

The mouth is located at or near the anterior end and encircled by tentacles, which are used to pull food into the mouth. They are buccal podia (i.e. related to the mouth cavity) and can be extended by hydraulic pressure (Figs. 1.7–1.10). The tentacles gather the food, whether by filtering seawater or by sweeping the substrates (Bouland et al., 1982). When sea cucumbers feed, each tentacle is wiped off periodically inside the oesophagus to remove the food. The number of these tentacles differs among sea cucumber species (from 10 to 30 tentacles), and their shape can be dendritic, pinnate, digitate or peltate (Lambert, 1997; Hasbún and Lawrence, 2002; Burton and Burton, 2002; Massin et al., 2002; Piper, 2007).

Unlike other animals, sea cucumbers breathe from their anus through the respiratory tree, which serves as lungs, allowing gas diffusion (Spirina and Dolmatov, 2001). Numerous small animals can live inside the cloaca for protection or food (Eeckhaut et al., 2004). Still, some sea cucumber species, especially the genus *Actinopyga*, have five anal teeth presumably to keep away any unwanted visitors (Fig. 1.11).

FIGURE 1.7

The tentacles and the tube feet of sea cucumber *Pearsonothuria graeffei*.

Credit: By Frédéric Ducarme. CC BY-SA 4.0 via Wikimedia Commons.

FIGURE 1.8

The branched tentacles of the sea cucumber *Cucumaria miniata*.

Credit: Kelly Cunningham, attribution via Wikimedia Commons.

FIGURE 1.9

The tentacles of the sea cucumber *Apostichopus californicus*.

Credit: Drow male. CC BY-SA 4.0 via Wikimedia Commons.

FIGURE 1.10

The tentacles of the sea cucumber *Synapta maculata*.

Credit: Rickard Zerpe. CC-BY 2.0 via Wikimedia Commons.

FIGURE 1.11

The anal teeth in the sea cucumber *Actinopyga caerulea*.

Credit: Julien Bidet. CC BY-SA 4.0 via Wikimedia Commons.

1.5 Extraordinary animals

Sea cucumbers exhibit many fascinating characteristics that are rarely seen in other aquatic animals. Sea cucumbers are epibenthic, and they hide in shelters during the daytime and increase their activity at nighttime. They have neither brain nor eyes, but one mouth and one anus. They have an endoskeleton in the form of microscopic ossicles embedded in their body wall. Although epibenthic in their adult form, some sea cucumbers engage in a kind of swimming behaviour in some cases to avoid predators and physical hazards and to seek for feeding resources. Also, some holothurians displayed swimming behaviour for spreading and spawning (Margolin, 1976; McEuen, 1988; Rogacheva et al., 2012). Sea cucumbers can modulate and

FIGURE 1.12

The sea cucumber *Actinopyga agassizii* during spawning posture.

Credit: Public domain via Wikimedia Commons.

maintain their inner density to float in the water column (Fig. 1.12). Before swimming, benthopelagic sea cucumbers were observed emptying their intestines (i.e. defecating), which occupies a large part of the body volume. This behaviour is likely to achieve buoyancy (Rogacheva et al., 2012).

Furthermore, sea cucumbers with respiratory trees inside their bodies are nutritionally bipolar; the anus serves as a second mouth for ingesting particulates and dissolved nutrients (Jaeckle and Strathmann, 2013). Sea cucumbers, excluding members of the orders Apodida and Elasipodida, have branched tubes inside the body cavity called respiratory trees. These structures not only allow the exchange of gases between the water and body cavity of the sea cucumber but also assimilate macromolecules from the seawater and transport these nutrients to the haemal system (Jaeckle and Strathmann, 2013).

Additionally, sea cucumbers can expel their internal organs as a defensive strategy to keep predators at bay, and they exhibit an exceptional regenerative ability to regenerate them. They can even reproduce asexually to a new adult at a higher rate than sea stars and sea urchins (Ortiz-Pineda et al., 2009; Zhang et al., 2017).

Moreover, sea cucumbers have been utilised from ancient times as food and traditional medicine (Chen, 2003). In addition to their beneficial influence on human health, they play essential roles in marine ecosystems. Their activities can improve sediment characteristics, enhance water chemistry, remove aquaculture wastes and influence the productivity of many benthic organisms (Purcell et al., 2016).

1.6 Treasure of the seabed

Sea cucumbers have been used for food and traditional medicine since ancient times (Fig. 1.13). In recent decades, sea cucumbers have gained much attention from researchers worldwide because of their high-valued compounds, in an attempt to convince customers of their beneficial influence on human health and possible therapy. Sea cucumbers are consumed mainly for their body wall; sometimes, the gonads and intestines are consumed because of their nutritional values (Purcell et al., 2012; Yuan et al., 2010). Sea cucumber contains many high-valued compounds such as

FIGURE 1.13

Dried sea cucumbers in a pharmacy.

Credit: By Chris 73. CC BY-SA 3.0 via Wikimedia Commons.

bioactive peptides, carotenoids, triterpene glycosides, minerals (for instance, zinc, calcium, iron, magnesium), vitamins (for instance, A, B_1, B_2, B_3), fatty acids, protein and polysaccharides, which make sea cucumber very attractive for consumption (Zhao et al., 2008).

Sea cucumbers contain high protein value for nutrition with a positive influence on the serum triglyceride (Taboada et al., 2003). The protein content of sea cucumbers can reach as high as 40.7%−63.3% (Omran, 2013; Wen et al., 2010). The primary amino acids in sea cucumbers are glycine, glutamic acid, aspartic acid, alanine and arginine (Omran, 2013; Wen et al., 2010; Zhong et al., 2007). Glycine from sea cucumbers can enhance immunity. Glycine and glutamic acids are essential for producing glutathione that initiates activation and creation of neutral killer cells (Zhao et al., 2008). Consuming the body wall of the sea cucumber can stimulate the immunity function as it is rich in arginine, glycine and glutamic acid (Zhao et al., 2008). Moreover, consuming 3 g of the dried body wall per day is for reducing arthralgia (Chen, 2003).

The fat content in sea cucumbers can be as low as 0.3%−5.66% (Omran, 2013; Wen et al., 2010). Sea cucumbers contain unsaturated fatty acids, such as omega-3 and omega-6 fatty acids, which are unique components of the body wall. They are beneficial against heart diseases (Harper and Jacobson, 2005). The common fatty acids in the sea cucumber are palmitic acid (C16:0), arachidonic acid (C20:4n6) and eicosapentaenoic acid (EPA, C20:5n−3), ranging from 43.2% to 56.7% of the total fatty acids (Zhong et al., 2007; Omran, 2013; Yahyavi et al., 2012). The long-chain fatty acids that are detected from sea cucumbers, mainly arachidonic acid, play a role in wound healing (Drazen et al., 2008; Fredalina et al., 1999; Jais et al., 1994; Svetashev et al., 1991).

Polysaccharides extracted from the body wall of the sea cucumber have anti-coagulant activities (Luo et al., 2013). Three types of polysaccharides are found in the body wall of sea cucumbers: neutral glycans, fucosylated chondroitin sulphates and sulphated fucans (Luo et al., 2013; Mulloy et al., 2000; Pomin, 2014).

In addition to their nutritional values, sea cucumbers contain numerous bioactive compounds such as chondroitin sulphate, triterpene glycosides saponins, lectins, heparin, cerebrosides, gangliosides, bioactive peptides, sterols and omega-6 and omega-3 fatty acids; these compounds are linked to multiple biologic and pharmacologic activities, such as anti-angiogenic, anti-cancer, anti-coagulant, anti-hypertension, anti-inflammatory, anti-microbial, anti-fungal, anti-oxidant, anti-thrombotic and anti-tumour (Fig. 1.14) (Table 1.1). Also, sea cucumbers are believed to possess a wide range of health functions. Sea cucumbers nourish the body, tonify the kidney and moisten the intestines. Also, sea cucumbers are known to treat stomach ulcers, asthma, hypertension, rheumatism, impotence and wounds. Furthermore, sea cucumbers regulate heart, kidney and lung activities and promote spermatogenesis (Bordbar et al., 2011; Borsig et al., 2007; Chen, 2003; Fredalina et al., 1999; Janakiram et al., 2015; Khotimchenko, 2018; Kiew and Don, 2012; Ming, 2001; Mourao et al., 1996; Pangestuti and Arifin, 2018).

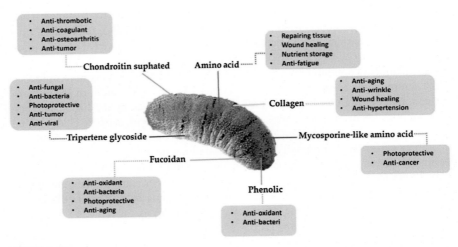

FIGURE 1.14

Bioactive compounds of sea cucumbers and their beneficial actions on human health.

Credit: Siahaan, E.A., Pangestuti, R., Munandar, H. and Kim, S.K., 2017. Cosmeceuticals properties of sea cucumbers: prospects and trends. Cosmetics 4 (3), 26. Licensed under (CC BY 4.0).

Table 1.1 Pharmacologic and medicinal activities of bioactive compounds from sea cucumbers.

Sea cucumber species	Bioactive compounds	Pharmacologic/medicinal activity	References
Pentacta quadrangularis, *Cucumaria frondosa*	Sulphated saponin (philinospide A), philinospide E, sea cucumber fractions: B1000 and fFucosylated chondroitin sulphate	Anti-angiogenic	Tian et al. (2005), Tong et al. (2005), Collin (1999)
Holothuria scabra, *Holothuria leucospilota*, *Stichopus chloronotus*, *C. frondosa*, *Cucumaria okhotensis*, *Mensamaria intercedens*, *Pearsonothuria graeffei*, *Apostichopus japonicus*, *Stichopus variegatus*	Triterpenoid (Frondoside A), triterpene oligoglycosides (okhotosides B1, B2 and B3), triterpene glycosides (intercedensides A, B and C), glycolipid (frondanol A), triterpene oligoglycosides (holothurin A and 24-dehydroechinoside), frondanol(R)-A5p, sphingoid base composition of cerebrosides	Anti-cancer	Althunibat et al. (2009), Zou et al. (2003), Sugawara et al. (2006), Janakiram et al. (2010), Ogushi et al. (2005), Li et al. (2008), Zhao et al. (2010)

Table 1.1 Pharmacologic and medicinal activities of bioactive compounds from sea cucumbers.—*cont'd*

Sea cucumber species	Bioactive compounds	Pharmacologic/ medicinal activity	References
Ludwigothurea grisea, Thelenota ananas	Fucosylated chondroitin sulphate, Fucosylated chondroitin sulphate	Anti-coagulant	Mulloy et al. (2000), Wu et al. (2010), Mourao et al. (1996)
A. japonicus	Low-molecular-weight polypeptides, polypeptides, acidic mucopolysaccharides, collagen and bioactive amino acids (all together)	Anti-fatigue	Fu and Cui (2007), Bing et al. (2010), Wang et al. (2007), Liu et al. (2009)
Actinopyga echinites, Actinopyga miliaris, Holothuria atra, H. scabra, Bohadschia argus, C. frondosa, Holothuria poli, Hemoiedema spectabilis, Psolus patagonicus, Actinopyga lecanora, Bohadschia marmorata, C. frondosa	Steroidal sapogenins, (Phosphate-buffered saline), sulphated triterpene glycosides (hemoiedemosides A and B), triterpene glycoside (patagonicoside A), triterpene glycoside (holothurin B [saponin]), triterpene glycoside (patagonicoside A), holostan-type triterpene glycosides (marmoratoside A, 17α-hydroxy impatienside A and bivittoside D), bioactive peptides	Anti-microbial Anti-bacterial and anti-fungal	Beauregard et al. (2001), Murray et al. (2001), Jawahar et al. (2002), Haug et al. (2002), Kumar et al. (2007), Farouk et al. (2007), Ismail et al. (2008), Chludil et al. (2002), Muniai et al. (2008), Yuan et al. (2009), Ridzwan et al. (1995)
C. frondosa, A. japonicus, Paracaudina chilensis, H. scabra, H. leucospilota, S. chloronotus, Acaudina molpadioides	Gelatin hydrolysate, gelatin hydrolysate, protein hydrolysate (bioactive peptides), bioactive peptides, phenols and flavonoids, gelatin hydrolysate (bioactive peptides), collagen polypeptides	Anti-oxidation	Zhong et al. (2007), Zeng et al. (2007), Mamelona et al. (2007, 2010), Su et al. (2009), Wang et al. (2010), Huihui et al. (2010), Althunibat et al. (2009)

Continued

Table 1.1 Pharmacologic and medicinal activities of bioactive compounds from sea cucumbers.—*cont'd*

Sea cucumber species	Bioactive compounds	Pharmacologic/ medicinal activity	References
A. japonicus	Glycosaminoglycan, holothurian glycosaminoglycan	Anti-thrombotic	Borsing et al. (2007), Zancan and Mourão (2004), Suzuki et al. (1991), Li et al. (2000)
M. intercedens, Holothuria hilla, P. quadrangularis, Holothuria forskali, A. japonicus, Holothuria nobilis, Holothuria fuscocinerea, Holothuria impatiens, L. grisea, Cucumaria japonica	Triterpene glycosides, (intercedensides D—I), glycoprotein (GPMI I), triterpene glycosides (hillasides A and B), sulphated saponins (philinopside A), triterpene glycosides (holothurinosides A, B, C and D and desholothurin A), mucopolysaccharide (SJAMP), triterpene glycosides (nobilisides A, B and C), triterpene glycosides (fuscocinerosides A, B and C), monosulphated triterpene glycosides, anostane-type triterpene glycoside (impatienside A), sulphated polysaccharide, monosulphated triterpene glycosides (cumaside)	Anti-tumour	Zou et al., (2003), Tong et al. (2005), Aminin et al. (2010), Wu et al. (2000), Lu et al. (2009), Rodriguez et al. (1991), Zhang and Yin (2010), Ogushi et al. (2006), Wu et al. (2007), Sun et al. (2007)
Staurocucumis liouvillei	Trisulphated triterpene glycosides (liouvillosides A and B)	Anti-viral	Maier et al. (2001)
A. japonicus	Fucan sulphate, glucosamine, chondroitin	Osteoarthritis	Kariya et al. (2004)

Table 1.1 Pharmacologic and medicinal activities of bioactive compounds from sea cucumbers.—*cont'd*

Sea cucumber species	Bioactive compounds	Pharmacologic/ medicinal activity	References
Thyone briareus, S. chloronotus, Stichopus hermanni, T. ananas, Thelenota anax, Holothuria fuscogilva, Holothuria fuscopunctata, Actinopyga mauritiana, Actinopyga caerulea, B. argus, S. chloronotus, Holothuria tubulosa, H. poli, Holothuria mammata	Polyunsaturated fatty acids, (arachidonic acid, eicosapentaenoic acid, docosahexaenoic acid)	Wound healing	Aydin et al. (2011), Wen et al. (2010), Fredalina et al. (1999), Svetashev et al. (1991), Drazen et al. (2008), Yaacob et al. (1994)

From Bordbar, S., Anwar, F., Saari, N., 2011. High-value components and bioactives from sea cucumbers for functional foods—a review. Mar. Drugs 9(10), 1761—1805. Licensed under CC BY 4.0.

References

Althunibat, O.Y., Hashim, R.B., Taher, M., Daud, J.M., Ikeda, M.A., Zali, B.I., 2009. In vitro anti-oxidant and antiproliferative activities of three Malaysian sea cucumber species. Eur. J. Sci. Res. 37 (3), 376—387.

Aminin, D.L., Chaykina, E.L., Agafonova, I.G., Avilov, S.A., Kalinin, V.I., Stonik, V.A., 2010. Antitumor activity of the immunomodulatory lead Cumaside. Int. Immunopharm. 10 (6), 648—654.

Aydın, M., Sevgili, H., Tufan, B., Emre, Y., Köse, S., 2011. Proximate composition and fatty acid profile of three different fresh and dried commercial sea cucumbers from Turkey. Int. J. Food Sci. Technol. 46 (3), 500—508.

Beauregard, K.A., Truong, N.T., Zhang, H., Lin, W., Beck, G., 2001. The detection and isolation of a novel anti-microbial peptide from the echinoderm *Cucumaria frondosa*. In: Phylogenetic Perspectives on the Vertebrate Immune System. Springer, Boston, MA, pp. 55—62.

Bing, L., Jing-feng, W., Jia, F., Xiao-lin, L., Hui, L., Qin, Z., Chang-hu, X., 2010. Antifatigue effect of sea cucumber *Stichopus japonicus* in mice. Food Sci. 2010 (31), 244—247.

Bordbar, S., Anwar, F., Saari, N., 2011. High-value components and bioactives from sea cucumbers for functional foods—a review. Mar. Drugs 9 (10), 1761—1805.

Borsig, L., Wang, L., Cavalcante, M.C., Cardilo-Reis, L., Ferreira, P.L., Mourão, P.A., Esko, J.D., Pavão, M.S., 2007. Selection blocking activity of a fucosylated chondroitin sulfate glycosaminoglycan from sea cucumber effect on tumor metastasis and neutrophil recruitment. J. Biol. Chem. 282 (20), 14984–14991.

Bouland, C., Massin, C., Jangoux, M., 1982. The fine structure of the buccal tentacles of *Holothuria forskali* (Echinodermata, Holothuroidea). Zoomorphology 101 (2), 133–149.

Burton, M., Burton, R., 2002. International Wildlife Encyclopedia.

Chang, Y., Shi, S., Zhao, C., Han, Z., 2011. Characteristics of papillae in wild, cultivated and hybrid sea cucumbers (*Apostichopus japonicus*). Afr. J. Biotechnol. 10 (63), 13780–13788.

Chen, J., 2003. Overview of sea cucumber farming and sea ranching practices in China. SPC beche-de-mer Inf. Bull. 18, 18–23.

Chludil, H.D., Muniain, C.C., Seldes, A.M., Maier, M.S., 2002. Cytotoxic and anti-fungal triterpene glycosides from the Patagonian sea cucumber *Hemoiedema s pectabilis*. J. Nat. Prod. 65 (6), 860–865.

Collin, P.D., 1999. Process for Obtaining Medically Active Fractions from Sea Cucumbers. U.S. Patent 5,876,762.

Conand, 2006. In: Proceedings of the CITES Workshop on the Conservation of Sea Cucumbers in the Families Holothuriidae and Stichopodidae (NMFS-OPR-34), p. 33.

Díaz-Balzac, C.A., Abreu-Arbelo, J.E., García-Arrarás, J.E., 2010. Neuroanatomy of the tube feet and tentacles in *Holothuria glaberrima* (Holothuroidea, Echinodermata). Zoomorphology 129 (1), 33–43.

Drazen, J.C., Phleger, C.F., Guest, M.A., Nichols, P.D., 2008. Lipid, sterols and fatty acid composition of abyssal holothurians and ophiuroids from the North-East Pacific Ocean: food web implications. Comp. Biochem. Physiol. B Biochem. Mol. Biol. 151 (1), 79–87.

Eeckhaut, I., Parmentier, E., Becker, P., Gomez da Silva, S., Jangoux, M., 2004. Parasites and biotic diseases in field and cultivated sea cucumbers. In: Advances in Sea Cucumber Aquaculture and Management, pp. 311–325.

Farouk, A.E.A., Ghouse, F.A.H., Ridzwan, B.H., 2007. New bacterial species isolated from Malaysian sea cucumbers with optimised secreted antibacterial activity. Am. J. Biochem. Biotechnol. 3 (2), 60–65.

Fredalina, B.D., Ridzwan, B.H., Abidin, A.Z., Kaswandi, M.A., Zaiton, H., Zali, I., Kittakoop, P., Jais, A.M., 1999. Fatty acid compositions in local sea cucumber. Gen. Pharmacol. Vasc. Syst. 33 (4), 337–340.

Fu, X.J., Cui, Z.F., 2007. Anti-fatigue effects of lower polypeptide from sea cucumber on mice. Food Sci. Technol. 4, 259–261.

Harper, C.R., Jacobson, T.A., 2005. Usefulness of omega-3 fatty acids and the prevention of coronary heart disease. Am. J. Cardiol. 96 (11), 1521–1529.

Hasbún, C.R., Lawrence, A.J., 2002. An annotated description of shallow water holothurians (Echinodermata: Holothuroidea) from Cayos Cochinos, Honduras. Rev. Biol. Trop. 50 (2), 669–678.

Haug, T., Kjuul, A.K., Styrvold, O.B., Sandsdalen, E., Olsen, Ø.M., Stensvåg, K., 2002. Antibacterial activity in *Strongylocentrotus droebachiensis* (Echinoidea), *Cucumaria frondosa* (Holothuroidea), and *Asterias rubens* (Asteroidea). J. Invertebr. Pathol. 81 (2), 94–102.

Huihui, C., Ping, Y., Jianrong, L., 2010. The preparation of collagen polypeptide with free radical scavenging ability purified from *Acaudina molpadioides* Semper. J. Chin. Inst. Food Sci. Technol. 1.

Ismail, H., Lemriss, S., Aoun, Z.B., Mhadhebi, L., Dellai, A., Kacem, Y., Boiron, P., Bouraoui, A., 2008. Anti-fungal activity of aqueous and methanolic extracts from the Mediterranean sea cucumber, *Holothuria polii*. J. Mycolog. Med. 18 (1), 23–26.

Jaeckle, W.B., Strathmann, R.R., 2013. The anus as a second mouth: anal suspension feeding by an oral deposit-feeding sea cucumber. Invertebr. Biol. 132 (1), 62–68.

Jais, A.M.M., McCulloch, R., Croft, K., 1994. Fatty acid and amino acid composition in haruan as a potential role in wound healing. Gen. Pharmacol. Vasc. Syst. 25 (5), 947–950.

Janakiram, N.B., Mohammed, A., Zhang, Y., Choi, C.I., Woodward, C., Collin, P., Steele, V.E., Rao, C.V., 2010. Chemopreventive effects of frondanol A5, a *Cucumaria frondosa* extract, against rat colon carcinogenesis and inhibition of human colon cancer cell growth. Cancer Prev. Res. 3 (1), 82–91.

Janakiram, N.B., Mohammed, A., Rao, C.V., 2015. Sea cucumbers metabolites as potent anti-cancer agents. Mar. Drugs 13 (5), 2909–2923.

Jawahar, A.T., Nagarajan, J., Shanmugam, S.A., 2002. Anti-microbial substances of potential biomedical importance from holothurian species. Indian J. Mar. Sci. 31, 161–164.

Kariya, Y., Mulloy, B., Imai, K., Tominaga, A., Kaneko, T., Asari, A., Suzuki, K., Masuda, H., Kyogashima, M., Ishii, T., 2004. Isolation and partial characterization of fucan sulfates from the body wall of sea cucumber *Stichopus japonicus* and their ability to inhibit osteoclastogenesis. Carbohydr. Res. 339 (7), 1339–1346.

Khotimchenko, Y., 2018. Pharmacological potential of sea cucumbers. Int. J. Mol. Sci. 19 (5), 1342.

Kiew, P.L., Don, M.M., 2012. Jewel of the seabed: sea cucumbers as nutritional and drug candidates. Int. J. Food Sci. Nutr. 63 (5), 616–636.

Kumar, R., Chaturvedi, A.K., Shukla, P.K., Lakshmi, V., 2007. Anti-fungal activity in triterpene glycosides from the sea cucumber *Actinopyga lecanora*. Bioorg. Med. Chem. Lett 17 (15), 4387–4391.

Lambert, P., 1997. Sea Cucumbers of British Columbia, Southeast Alaska and Puget Sound.

Li, Z., Wang, H., Li, J., Zhang, G., Gao, C., 2000. Basic and clinical study on the anti-thrombotic mechanism of glycosaminoglycan extracted from sea cucumber. Chin. Med. J. 113, 706–711.

Li, X., Roginsky, A.B., Ding, X.Z., Woodward, C., Collin, P., Newman, R.A., Bell Jr., R.H., Adrian, T.E., 2008. Review of the apoptosis pathways in pancreatic cancer and the anti-apoptotic effects of the novel sea cucumber compound, frondoside A. Ann. N. Y. Acad. Sci. 1138 (1), 181–198.

Liu, C., Wang, X., Yuan, W., Meng, X., Xia, X., Zhang, M., Tang, J., Hu, W., Sun, Y., Liu, J., 2009. Anti-fatigue and immune functions of sea cucumber oral liquid. Mod. Food Sci. Tech. 25 (10), 1115–1119.

Lu, Y.T., Kang, C.M., Xue, C.H., 2009. Separation and purification of cerebrosides from sea cucumbers. Food Sci. http://en.cnki.com.cn/Article_en/CJFDTOTAL-SPKX200911006.htm. (Accessed August 2020).

Luo, L., Wu, M., Xu, L., Lian, W., Xiang, J., Lu, F., Gao, N., Xiao, C., Wang, S., Zhao, J., 2013. Comparison of physicochemical characteristics and anti-coagulant activities of polysaccharides from three sea cucumbers. Mar. Drugs 11 (2), 399–417.

Maier, M.S., Roccatagliata, A.J., Kuriss, A., Chludil, H., Seldes, A.M., Pujol, C.A., Damonte, E.B., 2001. Two new cytotoxic and virucidal trisulfated triterpene glycosides from the Antarctic sea cucumber *Staurocucumis liouvillei*. J. Nat. Prod. 64 (6), 732–736.

Mamelona, J., Pelletier, E., Girard-Lalancette, K., Legault, J., Karboune, S., Kermasha, S., 2007. Quantification of phenolic contents and anti-oxidant capacity of Atlantic sea cucumber, *Cucumaria frondosa*. Food Chem. 104 (3), 1040–1047.

Mamelona, J., Saint-Louis, R., Pelletier, É., 2010. Nutritional composition and anti-oxidant properties of protein hydrolysates prepared from echinoderm byproducts. Int. J. Food Sci. Technol. 45 (1), 147–154.

Margolin, A.S., 1976. Swimming of the sea cucumber *Parastichopus californicus* (Stimpson) in response to sea stars. Ophelia 15 (2), 105—114.

Massin, C., Zulfigar, Y., Hwai, T.S., Boss, S.R., 2002. The genus Stichopus (Echinodermata: Holothuroidea) from the Johore Marine Park (Malaysia) with the description of two new species. Bull. Inst. Roy. Sci. Nat. Belgique 72, 73—99.

McEuen, F.S., 1988. Spawning behaviors of Northeast Pacific sea cucumbers (Holothuroidea: Echinodermata). Mar. Biol. 98 (4), 565—585.

Miller, A.K., Kerr, A.M., Paulay, G., Reich, M., Wilson, N.G., Carvajal, J.I., Rouse, G.W., 2017. Molecular phylogeny of extant Holothuroidea (Echinodermata). Mol. Phylogenet. Evol. 111, 110—131.

Ming, S.H.E.N., 2001. Investigation on component and pharmacology of sea cucumber. Chin. Tradit. Pat. Med. 10, 21.

Motokawa, T., Keegan, B.F., O'Connor, B.D., 1985. Catch connective tissue: the connective tissue with adjustable mechanical properties. Echinodermata 69—73.

Mourão, P.A., Pereira, M.S., Pavão, M.S., Mulloy, B., Tollefsen, D.M., Mowinckel, M.C., Abildgaard, U., 1996. Structure and anti-coagulant activity of a fucosylated chondroitin sulfate from echinoderm sulfated fucose branches on the polysaccharide account for its high anti-coagulant action. J. Biol. Chem. 271 (39), 23973—23984.

Mulloy, B., Mourão, P.A.S., Gray, E., 2000. Structure/function studies of anti-coagulant sulphated polysaccharides using NMR. J. Biotechnol. 77, 123—135. https://doi.org/10.1016/S0168-1656(99)00211-4.

Muniain, C., Centurión, R., Careaga, V.P., Maier, M.S., 2008. Chemical ecology and bioactivity of triterpene glycosides from the sea cucumber *Psolus patagonicus* (Dendrochirotida: Psolidae). Marine Biological Association of the United Kingdom. J. Mar. Biol. Assoc. U. K. 88 (4), 817.

Murray, A.P., Muniain, C., Seldes, A.M., Maier, M.S., 2001. Patagonicoside A: a novel antifungal disulfated triterpene glycoside from the sea cucumber *Psolus patagonicus*. Tetrahedron 57 (47), 9563—9568.

Ogushi, M., Yoshie-stark, M., Suzuki, T., 2005. Cytostatic activity of hot water extracts from the sea cucumber in Caco-2. Food Sci. Technol. Res. 11, 202—206.

Ogushi, M., Yoshie-Stark, Y., Suzuki, T., 2006. Apoptosis-inducing activity of hot water extracts from the sea cucumber in human colon tumor cells. Food Sci. Technol. Res. 12 (4), 290—294.

Omran, N.E.S.E.S., 2013. Nutritional value of some Egyptian sea cucumbers. Afr. J. Biotechnol. 12 (35).

Ortiz-Pineda, P.A., Ramírez-Gómez, F., Pérez-Ortiz, J., González-Díaz, S., Santiago-De Jesús, F., Hernández-Pasos, J., Del Valle-Avila, C., Rojas-Cartagena, C., Suárez-Castillo, E.C., Tossas, K., Méndez-Merced, A.T., 2009. Gene expression profiling of intestinal regeneration in the sea cucumber. BMC Genom. 10 (1), 262.

Pangestuti, R., Arifin, Z., 2018. Medicinal and health benefit effects of functional sea cucumbers. J. Tradit. Complementary Med. 8 (3), 341—351.

Paulay, G., Hansson, H., 2013. Holothuroidea. World Register of Marine Species. http://www.marinespecies.org/aphia.php?p=taxdetails&id=731943. (Accessed December 2018).

Pawson, D.L., Barraclough-Fell, H., 1965. A Revised Classification of the Dendrochirote Holothurians.

Piper, R., 2007. Extraordinary Animals: An Encyclopedia of Curious and Unusual Animals, vol. 125. Greenwood Press, London.

Pomin, V.H., 2014. Holothurian fucosylated chondroitin sulfate. Mar. Drugs 12 (1), 232—254.

Purcell, S.W., Samyn, Y., Conand, C., 2012. Commercially Important Sea Cucumbers of the World.

Purcell, S.W., Conand, C., Uthicke, S., Byrne, M., 2016. Ecological roles of exploited sea cucumbers. In: Oceanography and Marine Biology. CRC Press, pp. 375–394.

Ridzwan, B.H., Kaswandi, M.A., Azman, Y., Fuad, M., 1995. Screening for antibacterial agents in three species of sea cucumbers from coastal areas of Sabah. Gen. Pharmacol. 26 (7), 1539–1543.

Rodriguez, J., Castro, R., Riguera, R., 1991. Holothurinosides: new anti-tumour non sulphated triterpenoid glycosides from the sea cucumber *Holothuria forskalii*. Tetrahedron 47 (26), 4753–4762.

Rogacheva, A., Gebruk, A., Alt, C.H., 2012. Swimming deep-sea holothurians (Echinodermata: Holothuroidea) on the northern mid-Atlantic ridge. Zoosymposia 7 (1), 213–224.

Siahaan, E.A., Pangestuti, R., Munandar, H., Kim, S.K., 2017. Cosmeceuticals properties of sea cucumbers: prospects and trends. Cosmetics 4 (3), 26.

Smirnov, A.V., 2012. System of the class Holothuroidea. Paleontol 46, 793–832. https://doi.org/10.1134/S0031030112080126.

Spirina, I.S., Dolmatov, I.Y., 2001. Morphology of the respiratory trees in the holothurians *Apostichopus japonicus* and *Cucumaria japonica*. Russ. J. Mar. Biol. 27 (6), 367–375.

Su, Y.C., Liu, S.J., Wu, C.Y., 2009. Optimisation of the preparation procedure and the antioxidant activity of polypeptide from sea cucumber. J. Fujian Fish. 2, 6–10.

Sugawara, T., Zaima, N., Yamamoto, A., Sakai, S., Noguchi, R., Hirata, T., 2006. Isolation of sphingoid bases of sea cucumber cerebrosides and their cytotoxicity against human colon cancer cells. Biosci. Biotechnol. Biochem. 70, 2906–2912, 0611080146.

Sun, P., Liu, B.S., Yi, Y.H., Li, L., Gui, M., Tang, H.F., Zhang, D.Z., Zhang, S.L., 2007. A new cytotoxic lanostane-type triterpene glycoside from the sea cucumber *Holothuria impatiens*. Chem. Biodivers. 4 (3), 450–457.

Suzuki, N., Kitazato, K., Takamatsu, J., Saito, H., 1991. Anti-thrombotic and anti-coagulant activity of depolymerised fragment of the glycosaminoglycan extracted from *Stichopus japonicus* Selenka. Thromb. Haemostasis 66 (04), 369–373.

Svetashev, V.I., Levin, V.S., Lam, C.N., 1991. Lipid and fatty acid composition of holothurians from tropical and temperate waters. Comp. Biochem. Physiol. B 98 (4), 489–494.

Taboada, M.C., Gonzalez, M., Rodriguez, E., 2003. Value and effects on digestive enzymes and serum lipids of the marine invertebrate *Holothuria forskali*. Nutr. Res. 23 (12), 1661–1670.

Tian, F., Zhang, X., Tong, Y., Yi, Y., Zhang, S., Li, L., Sun, P., Lin, L., Ding, J., 2005. PE, a new sulfated saponin from sea cucumber, exhibits anti-angiogenic and anti-tumor activities in vitro and in vivo. Cancer Biol. Ther. 4 (8), 874–882.

Tong, Y., Zhang, X., Tian, F., Yi, Y., Xu, Q., Li, L., Tong, L., Lin, L., Ding, J., 2005. Philinopside a, a novel marine-derived compound possessing dual anti-angiogenic and anti-tumor effects. Int. J. Cancer 114 (6), 843–853.

Wang, H.-T., Yin, H.-X., Jin, H.-Z., Ha, J.-Y., 2007. The study of anti-fatigue effects of sea cucumber polypeptide on mice [J]. Food & Machinery 3. http://en.cnki.com.cn/Article_en/CJFDTOTAL-SPJX200703028.htm.

Wang, J., Wang, Y., Tang, Q., Wang, Y., Chang, Y., Zhao, Q., Xue, C., 2010. Antioxidation activities of low-molecular-weight gelatin hydrolysate isolated from the sea cucumber *Stichopus japonicus*. J. Ocean Univ. China 9 (1), 94–98.

Wen, J., Hu, C., Fan, S., 2010. Chemical composition and nutritional quality of sea cucumbers. J. Sci. Food Agric. 90 (14), 2469–2474.

Woo, S.P., Yasin, Z., Tan, S.H., Kajihara, H., Fujita, T., 2015. Sea cucumbers of the genus Stichopus Brandt, 1835 (Holothuroidea, Stichopodidae) in straits of malacca with description of a new species. ZooKeys (545), 1.

Wu, P., Chen, Y., Fang, J., Su, W., 2000. Studies on the chemical constituents from sea cucumber *Mensamaria intercedens* IV isolation, properties and antitumor activity of the glycoprotein from *Mensamaria intercedens*. Chin. J. Mar. Drugs. http://en.cnki.com.cn/Article_en/CJFDTOTAL-HYYW200005001.htm. (Accessed August 2020).

Wu, J., Yi, Y.H., Tang, H., Wu, H.M., Zhou, Z.R., 2007. Hillasides A and B, two new cytotoxic triterpene glycosides from the sea cucumber Holothuria hilla lesson. J. Asian Nat. Prod. Res. 9, 609—615.

Wu, M., Xu, S., Zhao, J., Kang, H., Ding, H., 2010. Free-radical depolymerisation of glycosaminoglycan from sea cucumber Thelenata ananas by hydrogen peroxide and copper ions. Carbohydr. Polym. 80 (4), 1116—1124.

Yaacob, H., Kim, K.H., Shahimi, M.M., Jamalullail, S.M.S., 1994. Water Extract of Stichopus SP1 (Gamat) Improves Wound Healing.

Yahyavi, M., Afkhami, M., Javadi, A., Ehsanpour, M., Khazaali, A., Mokhlesi, A., 2012. Fatty acid composition in two sea cucumber species, *Holothuria scabra* and *Holothuria leucospilata* from Qeshm Island (Persian Gulf). Afr. J. Biotechnol. 11 (12).

Yuan, W.H., Yi, Y., Tang, H.F., Liu, B.S., Wang, Z.L., Sun, G.Q., Zhang, W., Li, L., Sun, P., 2009. Anti-fungal triterpene glycosides from the sea cucumber *Bohadschia marmorata*. Planta Med. 75 (02), 168—173.

Yuan, W.P., Liu, C.H., Wang, X.J., Meng, X.M., Xia, X.K., Zhang, M.S., Hu, W., 2010. Evaluation and analysis of nutritional composition of different parts of sea cucumber *Apostichopus joponicus*. Sci. Technol. Food Ind. http://en.cnki.com.cn/Article_en/CJFDTOTAL-SPKJ201005091.htm. (Accessed August 2020).

Zancan, P., Mourão, P.A., 2004. Venous and arterial thrombosis in rat models: dissociation of the anti-thrombotic effects of glycosaminoglycans. Blood Coagulation Fibrinolysis 15 (1), 45—54.

Zeng, M., Xiao, F., Zhao, Y., Liu, Z., Li, B., Dong, S., 2007. Study on the free radical scavenging activity of sea cucumber (*Paracaudina chinens* var.) gelatin hydrolysate. J. Ocean Univ. China 6 (3), 255—258.

Zhang, X., Sun, L., Yuan, J., Sun, Y., Gao, Y., Zhang, L., Li, S., Dai, H., Hamel, J.F., Liu, C., Yu, Y., 2017. The sea cucumber genome provides insights into morphological evolution and visceral regeneration. PLoS Biol. 15 (10), e2003790.

Zhang, W.-W., Yin, L., 2010. Study advance of Holothuria in antitumor activities. China J. Tradit. Chin. Med. Pharm. http://en.cnki.com.cn/Article_en/CJFDTOTAL-BXYY201001035.htm.

Zhao, Q., Wang, J.F., Xue, Y., Wang, Y., Gao, S., Lei, M., Xue, C.H., 2008. Comparative study on the bioactive components and immune function of three species of sea cucumber. J. Fish. Sci. China 15 (1), 154—159.

Zhao, Q., Xue, Y., Liu, Z.D., Li, H., Wang, J.F., Li, Z.J., Wang, Y.M., Dong, P., Xue, C.H., 2010. Differential effects of sulfated triterpene glycosides, holothurin A1, and 24-dehydroechinoside A, on antimetastasic activity via regulation of the MMP-9 signal pathway. J. Food Sci. 75 (9), H280—H288.

Zhong, Y., Khan, M.A., Shahidi, F., 2007. Compositional characteristics and anti-oxidant properties of fresh and processed sea cucumber (*Cucumaria frondosa*). J. Agric. Food Chem. 55 (4), 1188—1192.

Zou, Z.R., Yi, Y.H., Wu, H.M., Wu, J.H., Liaw, C.C., Lee, K.H., 2003. Intercedensides A—C, three new cytotoxic triterpene glycosides from the sea cucumber Mensamaria intercedens Lampert. J. Nat. Prod. 66 (8), 1055—1060.

Anatomic structure and function

2.1 Body walls

The body wall of sea cucumbers has unique properties, which make it attractive for consumption. The processed body wall of sea cucumbers is famous by trepang, bêche-de-mer, namako, haishen, plingkao or hai-som (Truong and Le, 2019). On the inner surface of the body wall, there are five longitudinal and radical muscle lines. The five longitudinal muscle bands are medially attached by connective tissue to the undersurface (Fig. 2.1) (Dolmatov et al., 1996). The body wall of sea cucumbers consists of epidermis and dermis layers. The epidermis layer constitutes long and conical cells covered by a thin cuticle (Guerrero Guerrero and Rodríguez Forero, 2018). The dermis layer constitutes a large part of the body wall and consists of an outer subepidermal layer, a middle dense connective tissue layer and an inner loose connective tissue (Byrne, 2001). The ossicles can be found in the outer layer below the epidermis (Woo et al., 2015). Pigment granules, in which the body gains its colouration, occur as free granules in the body walls and are often associated with nervous elements of the dermis (Hyman, 1955). The morula cells, which are possibly neurosecretory, are often found in the dermis layer (Motokawa, 1984).

The connective tissue of the sea cucumber is collagenous connective tissue. This tissue has a changeable mechanical property controlled by nerves. When sea cucumbers are touched, they stiffen their body walls. Also, they control their movement by softening and stiffening the body wall (Motokawa, 1984). This property allows the control of the cloacal pump. The muscle of the body wall pushes the coelomic fluid back and forth in a sluggish manner of circulation (Lambert, 1997). Also, sea cucumbers can extend their bodies longer than their original length. These mechanical properties allow control in the body tone in response to chemical, mechanical, photic and electrical exposure (Motokawa, 1984).

2.2 Ossicles and calcareous ring

Ossicles are microscopic and calcified elements that serve as an endoskeleton (Fig. 2.2). They can be found in the tentacle, body walls, papillae, podia and in the internal organs of sea cucumbers (Kamarudin and Rehan, 2015). The ossicles

Sea Cucumbers. https://doi.org/10.1016/B978-0-12-824377-0.00006-2

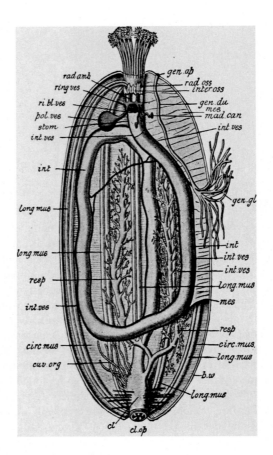

FIGURE 2.1

Internal organs of a holothurian. *b.w.*, body wall; *circ. mus.*, circular layer of muscle; *cl*, cloaca; *cl. op.*, cloacal opening with five teeth; *cuv. org.*, Cuvierian organs; *gen. ap.*, genital aperture; *gen. du*, genital duct; *gen. gl*, genital gland; *int.*, intestines; *inter. oss.*, interambulacral ossicles; *int. ves.*, intestinal vessels; *long. mus.*, longitudinal muscles; *mad. can.*, Madreporic canal; mes., mesentery; pol. ves., Polian vesicles; rad. amb., radial ambulacral vessel; *rad. oss.*, radial ossicles; *ri. bl. ves.*, ring blood vessel; *resp.*, respiratory trees; *ring ves.*, ring vessel of the ambulacral system; *stom*, stomach.

Credit: Public domain via Wikimedia commons.

serve as a critical taxonomic classification feature in the holothuroids because of their different shapes; the closely related species have similar shapes of ossicles (Lambert, 1997). The shapes of ossicles in the sea cucumber body include table, button, rod, anchor, perforated plate and rosettes (Fig. 2.3). These ossicles also differ among different parts of the same animal. For instance, Ahmed (2009) observed the ossicles in the body walls of the sea cucumber *Actinopyga mauritiana* from

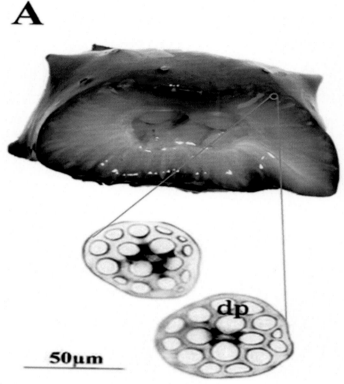

FIGURE 2.2

Ossicles in the sea cucumber *Apostichopus japonicus*.

Credit: Zhang, X., Sun, L., Yuan, J., Sun, Y., Gao, Y., Zhang, L., Li, S., Dai, H., Hamel, J.F., Liu, C., Yu, Y., 2017.
The sea cucumber genome provides insights into morphological evolution and visceral regeneration. PLoS Biol.,
15(10), e2003790. Licensed under CC BY 4.0.

the Red Sea coast of Egypt. The ossicles in the tentacles were found large and rugose rods; in the dorsal body walls, they were spiny rods and simple and tiny rosettes; and in the ventral body wall were small grains, elongated grains and rods that can be spiny or smooth (Ahmed, 2009).

The calcareous ring is a hard part that surrounds the pharynx (Figs. 2.1 and 2.4). It supports the pharynx, the nerve ring and the water vascular system. Also, it serves as a base for operating the tentacles and an attachment point for the anterior part of the longitudinal muscles. It is formed from ossicles that vary in shape and size between holothurians, and they are bound together by connective tissue (Reich, 2015). The shape of the ring is also used as a classification feature for sea cucumbers (Massin and Lane, 1991). For instance, sea cucumbers with a ring of long posterior tails are in the same family (Lambert, 1997). Furthermore, the calcareous ring can help to

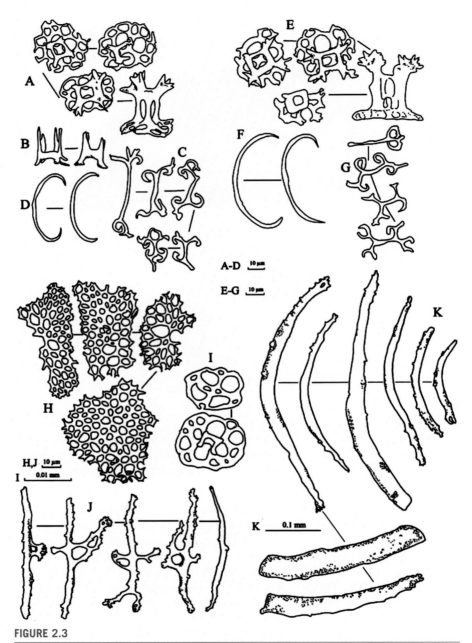

FIGURE 2.3

Spicules of *Stichopus herrmanni* Semper, 1868 (USM/MSL/PSEM004). (A) Tables from the dorsal body. (B) Pseudo tables from the dorsal body wall. (C) Rosettes from the dorsal body wall. (D) C-shaped rods from the dorsal body wall. (E) Tables from the dorsal

FIGURE 2.4

Calcareous ring detail of the sea cucumber *Cladolabes arafurus*.

Credit: Melanie MacKenzie. Licensed under CC BY 4.0. Via Wikimedia commons.

improve our knowledge of the early evolution of holothurians because it is the only part that fossilises (Lambert, 1997; Reich, 2015).

2.3 Circulatory system
2.3.1 Haemal system

Unlike other echinoderms, the haemal system in holothurians is more complicated. It consists of a central haemal ring around the pharynx that sends off sinuses, supplying the digestive tract, gonads, tentacles, water canals, podia and respiratory trees (Hyman, 1955; Ruppert and Barnes, 1994). The sinuses are associated with the gut and form ducts that run outside the digestive tract and lacunae in the connective tissue layer of the digestive wall. The wall of the haemal ducts consists of the outer muscular mesothelium associated with the nerve plexus and inner connective tissue layer. The nerve plexus associated with the haemal duct is isolated or runs in small bundles (Feral and Massin, 1982).

◀───────────────────────────────────────

papillae. (F) C-shaped rods from the dorsal papillae. (G) Rosettes from the dorsal papillae. (H) Large multiperforated plates from the tube feet. (I) Reduced tables from the tube feet. (J) Rods with central perforations from the tube feet. (K) Rods of different sizes from the tentacles.

Credit: Woo, S.P., Yasin, Z., Tan, S.H., Kajihara, H., Fujita, T., 2015. Sea cucumbers of the genus Stichopus Brandt, 1835 (Holothuroidea, Stichopodidae) in straits of malacca with description of a new species. ZooKeys 2015, 1–26. Licensed under CC BY 4.0.

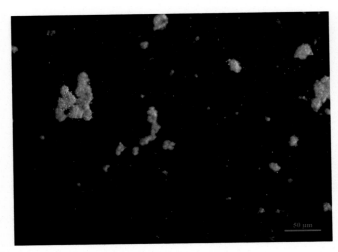

FIGURE 2.5

The coelomic fluid and the suspended coelomocytes in the sea cucumber *Apostichopus japonicus* under a fluorescence microscope. Scale bar: 50 μm.

Credit: Mohamed Mohsen.

2.3.2 Coelomic fluid and coelomocytes

The coelomic fluid fills the lumen inside the body and the inner viscera bathe in it. The coelomic fluid is similar, in some respects, to the blood. It fills the water vascular system and has many functions, such as hydraulic movement, transport of wastes and gases and storage of eggs and sperm during spawning. The coelomic fluid of sea cucumbers contains free cells called coelomocytes. They are suspended in the coelomic fluid and involved in immune defence, nutrient transport, gas exchange and waste execration (Fig. 2.5) (Eliseikina and Magarlamov, 2002). Holothurians contain five basic types of coelomocytes: haemocytes, phagocytes, spherule cells, lymphocytes and crystal cells (Hetzel, 1963). Among these types, phagocytes, lymphocytes and spherule cells are believed to exist in the coelomic fluid of all holothurians (Hetzel, 1963; Ramírez-Gómez et al., 2010).

2.4 Digestive system

2.4.1 Morphology

The digestive tract of sea cucumbers is a tubular duct. It begins with the mouth that is surrounded by modified tube feet, the tentacles. Most of the sea cucumbers have long and looped intestines (Vergara and Rodríguez, 2015). The digestive tract comprises the pharynx, oesophagus, stomach, intestines, rectum and cloaca. The pharynx is a short segment that extends from the mouth to the water vascular system or the calcareous ring. The pharynx is muscular in most of the holothurians and relatively thin-walled in Synaptidae, a family of sea cucumbers. The stomach is enlarged and has an anterior and a posterior contraction. The intestine can be divided into the anterior, middle and posterior intestine (Fig. 2.6) (Hyman, 1955; Jangoux and Lawrence, 1982). After dissection, the intestine is divided often into a first part descending and ascending and a second part descending (Kamenev et al., 2013). In most holothurians, the intestine is thick and ends in a swollen cloaca. The cloaca opens externally through the anus. The cloaca is short, thick and linked to the body walls by numerous suspensor strands (Fig. 2.1). The intestine is linked to the body wall by mesentery in some parts or suspended in the coelomic fluid (Hyman, 1955). In most of the sea cucumbers, the first part of the gut (oesophagus, stomach and the first descending segment of the intestine) is suspended by the mid-dorsal mesentery located in the interradius, and a left lateral mesentery suspends the ascending intestinal segment. The second descending intestinal segment is suspended by a right ventral mesentery (Feral and Massin, 1982). Also, in some holothurians, the anterior part of the digestive tract may be pigmented. The accumulated ingested sediment is subsequently ground in the pharynx, oesophagus and stomach. It is, after that, pushed into the intestines, wherein it is progressively compacted and coated with mucus, and then fragmented into faecal pellets, presumedly in the posterior section of the intestines (Feral and Massin, 1982).

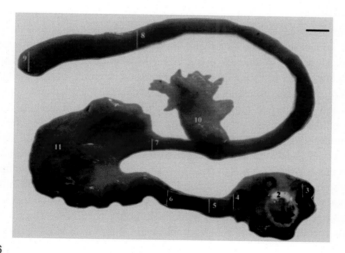

FIGURE 2.6

Gross anatomy of the sea cucumber *Isostichopus badionotus* intestine. 1: pharynx; 2: calcareous ring; 3: tentacles; 4: oesophagus; 5: stomach; 6: foregut; 7: midgut; 8: hindgut; 9: anus; 10: gonad; 11: mesenteric tissue. Scale: 1 cm.

Credit: Vergara, W., Rodríguez, A., 2015. Histology of the digestive tract of three species of sea cucumber Isostichopus badionotus, Stichopus *sp. and* Stichopus hermanni *(Aspidochirotida: Stichopodidae). Rev. Biol. Trop. 63(4), 1021-1033. Licensed under CC BY 4.0.*

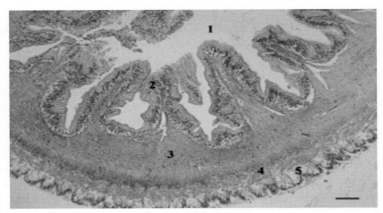

FIGURE 2.7

Overview of digestive tissue shows uniform villi separated by crypts situated on loose connective tissue. 1: intestinal lumen; 2: luminal epithelium; 3: connective tissue; 4: muscle tissue; 5: external epithelium or coelomic epithelium. HE Scale: 10 mm.

Credit: Vergara, W., Rodríguez, A., 2015. Histology of the digestive tract of three species of sea cucumber Isostichopus badionotus, Stichopus sp. and Stichopus hermanni *(Aspidochirotida: Stichopodidae). Rev. Biol. Trop. 63(4), 1021-1033. Licensed under CC BY 4.0. Licensed under CC BY 4.0.*

2.4.2 Histology

Starting from the digestive lumen, the wall of the intestine consists of the digestive epithelium, connective tissue layer and muscular mesothelium associated with a nerve plexus (Fig. 2.7) (Kamenev et al., 2013; Mashanov et al., 2004). Three basic cell types occur in the digestive tract of holothuroids: secretory cells, T-shaped cells and the enterocytes. The digestive epithelium consists of T-shaped cells, enterocytes, mucocytes and invading coelomocytes. The function of the T-shaped cells is the absorption of the dissolved organic matter, whereas the enterocytes are involved in intracellular digestion, absorption and lubrication of the gut wall. Mucocytes are also involved in lubrication of the gut wall (Feral and Massin, 1982). The coelomocytes cells form five types in the gut walls, mainly in the digestive epithelium and the connective tissue layer (Rosati, 1970). Feral and Massin (1982) described them as small coelomocytes, phagocytes and granulocytes. The connective tissue layer contains fibrocytes, coelomocytes cells and collagen fibres. The mesothelium muscle cells form circular and longitudinal layers (Feral and Massin, 1982).

2.5 Respiratory system

2.5.1 Morphology

Two branches of the respiratory tree extend in the body cavity of sea cucumbers from the orders Aspidochirota, Dendrochirota and Molpadonia (Hyman, 1955).

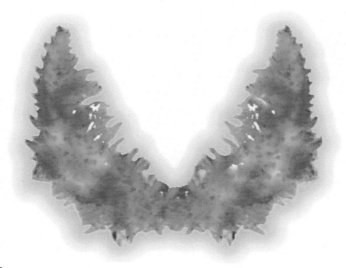

FIGURE 2.8

Drawing of the respiratory tree branches.

Credit: Mohamed Mohsen.

The respiratory trees are branched from the cloaca on either side of the intestine (Figs. 2.1 and 2.8).

The respiratory trees are considered as lungs for sea cucumbers that allow gas exchange. The respiratory tree opens independently by a duct into the cloaca. The sea cucumber draws in the oxygenated water through the anus by expanding the cloaca, allowing the gas exchange by diffusion through the thin membrane of the respiratory tree. Then, the water is driven outback by squeezing the body wall. Additionally, the body wall integument appears to account for approximately 40% of the total oxygen uptake in air-saturated water (Brown and Shick, 1979; Newell and Courtney, 1965).

2.5.2 Histology

The histology of the respiratory tree in the holothurian *Cladolabes schmeltzii* was reported as follows: the walls of the respiratory tree contain luminal epithelium, a connective tissue layer and an outer coelomic epithelium that runs in its central nerve plexus (Fig. 2.9). The cells of the luminal epithelium are irregular shaped and connected by intercellular junctions. The connective tissue forms a thick layer in the trunks, which becomes thinner in the branches of the respiratory tree. The coelomic epithelium consists of highly vacuolated peritoneocytes of irregular shape (Kamenev et al., 2013).

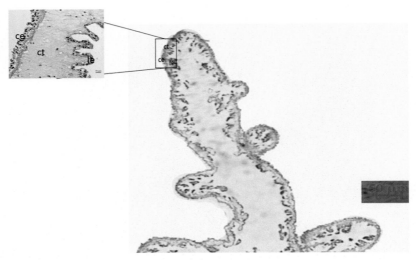

FIGURE 2.9

A histologic section in a branch of the respiratory tree of the sea cucumber *Apostichopus japonicus. ce,* coelomic epithelium; *ct,* connective tissue; *le,* luminal epithelium. HE Scale: 50 μm.

Credit: Mohamed Mohsen.

2.6 Water vascular system

The water vascular system in sea cucumbers is responsible for locomotion function by hydraulic pressure. The tentacles and the tube feet, which are filled with coelomic fluid, are controlled by the water vascular system. The tentacles are podia of the water vascular system (Hyman, 1955). The water vascular system consists of the following: (1) The ring canal surrounds the oesophagus and sends radial canal branches to the tentacles and then it curves back along the length of the body. It also extends to supply the tube feet by small branches. (2) Polian vesicles connect with the ring canal. The Polian vesicles may serve as a regulator for water pressure, and they have an excretory function (Baccetti and Rosati, 1968). (3) Madreporic canal below the pharynx filtrates the water that enters the water vascular system (Ezhova et al., 2017). In some holothurians, the madreporite performs an excretory function; it contains a pore that opens into the environment on the oral side under the tentacles. In other sea cucumbers, it losses connection with the outer environment and therefore connects the coelomic fluid with the water vascular system (Ezhova et al., 2017). Iwalaye et al. (2020) examined the pore sizes of the madreporite in the sea cucumber *Holothuria cinerascens*. The pore sizes of sea cucumbers (n = 10) ranged from 0.59 to 2.90 μm with the mean value of 1.22 ± 0.03 μm (Iwalaye et al., 2020).

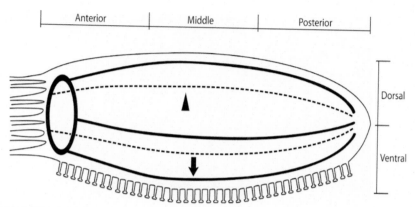

FIGURE 2.10

The holothurian nervous system starts with an anterior nerve ring from which five radial nerve cords extend up to the anus, which can be divided into ventral and dorsal according to the localisation of the tube feet (ventral).

Credit: Díaz-Balzac, C.A., Lázaro-Peña, M.I., Vázquez-Figueroa, L.D., Díaz-Balzac, R.J., García-Arrarás, J.E., 2016. Holothurian nervous system diversity revealed by neuroanatomical analysis: PloS One 11.

2.7 Nervous system

The nervous system of the holothurians starts with the nerve ring, which surrounds the pharynx ahead of the calcareous ring. Five radial nerve cords are extended from the nerve ring and divided into ventral and dorsal (Figs. 2.10 and 2.11) (Díaz-Balzac et al., 2010; Inoue et al., 1999; Mashanov et al., 2006). The radial nerve cord is subdivided into ectoneural and hyponeural regions by a connective tissue structure from which peripheral nerves extend to the other organs. Specifically, the peripheral nerves extending from the ectoneural nerve cord innervate the body wall, dermis, epidermis layers and tube feet. The peripheral nerves extending from the hyponeural nerve cord innervate the longitudinal muscle and the circular muscle (Díaz-Balzac et al, 2010, 2016; Inoue et al., 1999; Mashanov et al., 2006).

2.8 Reproductive system

Sea cucumbers mostly have separated sex, and the differentiation between males and females is difficult to identify from an external morphology. However, the sex in *Cucumaria frondosa* can be distinguished when the animal extends the tentacles in the water (Fig. 2.12) (Hamel and Mercier, 1996). The gonad in sea cucumbers constitutes a group of tubules that are gathered into a single duct. This duct opens outside through the gonopore on the dorsal side near to the tentacles. Most sea

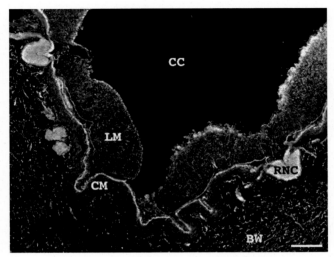

FIGURE 2.11

A transverse section of the nervous system of *Holothuria glaberrima* shows the ventral region radial nerves (RNC) and the nervous components of the body wall (BW), longitudinal muscle (LM), circular muscle (CM) and coelomic cavity (CC), which were differentiated by their immunoreactivity.

Credit: Díaz-Balzac, C.A., Lázaro-Peña, M.I., Vázquez-Figueroa, L.D., Díaz-Balzac, R.J., García-Arrarás, J.E., 2016. Holothurian nervous system diversity revealed by neuroanatomical analysis: PloS One 11.

FIGURE 2.12

The sea cucumber *Cucumaria frondosa*. Left, female, tube-shaped gonopore, one genital pore. Right, male, heart-shaped gonopore, numerous (5—12) papillae. Arrows point to the gonopore.

Credit: Dr Zonghe Yu, 2014.

Holothuria scabra versicolor

FIGURE 2.13

The posture of the sea cucumber during spawning.

Credit: Aymeric Desurmont. Gaudron, S., 2006. Observation of natural spawning of Bohadschia vitiensis. Beche-de-mer Inf. Bull. 24, 54.

cucumber species reproduce by distributing their sperm or egg into the water, whereby fertilisation occurs (Fig. 2.13).

Additionally, some sea cucumber species are capable of asexual production by fission (Dolmatov, 2014; Uthicke and Conand, 2005). Fission is the division of an organism into two parts or more, and those parts grow to independent organisms similar to the original one (Carlson, 2011). The simple example of the asexual reproduction is the division of the bacterium (Fig. 2.14), which can be described as follows: (1) The bacterium before binary fission has the DNA tightly coiled; (2) the DNA of the bacterium replicates; (3) the DNA moves to the two ends of the bacterium and the bacterium increases in size to prepare for splitting; (4) the equatorial plane of the cell constricts and separates the plasma membrane; (5) the growth of a new cell wall begins the separation of the bacterium; (6) the new cell wall fully

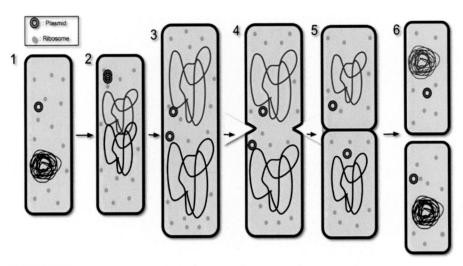

FIGURE 2.14

Binary fission in a prokaryote: an example.

Credit: Ecoddington14, 2014.

develops, resulting in the complete split of the bacterium, and finally, the new daughter cells have tightly coiled DNA, ribosomes and plasmids (Fig. 2.14).

In sea cucumbers, fission can happen using different strategies of stretching, twisting or constricting the body (Fig. 2.15). Specifically, the posterior section attaches a substrate by the tube feet and, simultaneously, the anterior part twists

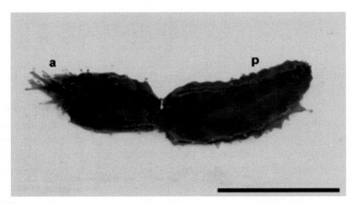

FIGURE 2.15

Twisting of the sea cucumber *Cladolabes schmeltzii* during fission. *a*, anterior part; *p*, posterior part. Scale bar: 2 cm.

Credit: Dolmatov, I.Y., 2014. Asexual reproduction in holothurians: Sci. World J. 2014. Licensed under CC BY 4.0.

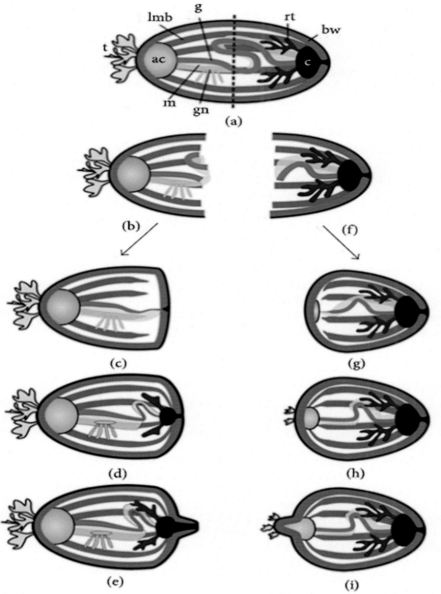

FIGURE 2.16

Scheme of regeneration of internal organs after fission in holothurians. (A) Animal before fission. (B) Anterior fragment just after fission. (C) Formation of gut and cloaca in the anterior fragment. (D) Formation of respiratory trees in the anterior fragment. (I) Growth of the posterior part of the body. (F) Posterior fragment just after fission. (G) Formation of ac and gut rudiments in the posterior fragment. (H) The posterior fragment with regenerated internal organs. (I) Growth of the anterior part of the body. *ac*, aquapharyngeal complex; *bw*, body wall; *c*, cloaca; *g*, gut; *gn*, gonad; *lmb*, longitudinal muscle band; *m*, mesentery; *rt*, respiratory tree; *t*, tentacles. *Dotted line*: site of the division of the body during fission.

or moves forward. During this stage, the body wall becomes soft in the part of fission preparing for this process (Motokawa, 1984). The duration of the fission varies between sea cucumber species from a few minutes to days, which likely depends on the intensity of the body wall. The sea cucumber can reproduce asexually in the adult stage and in the larvae stage. In the doliolaria stage of *Parastichopus californicus*, 12.2% of the cultured larvae undergo fission (Eaves and Richard Palmer, 2003). The ability of sea cucumbers to reproduce asexually may be used in nature to increase their natural population. The asexual reproduction by fission was recorded for many sea cucumber species; for instance, *Holothuria atra*, *Holothuria edulis*, *Holothuria leucospilota*, *Stichopus chloronotus*, *Stichopus horrens* and *Stichopus naso* (Purcell et al., 2012; Dolmatov, 2014).

The fission in sea cucumbers is accompanied by particular behaviour like twisting or stretching. Also, sea cucumbers stop feeding before fission. Furthermore, fission occurs in a specific area of the body, which is likely a species-specific area (Dolmatov, 2014). The regeneration of the internal organs is a process like an organ regeneration after evisceration in the sea cucumbers (Fig. 2.16). Various environmental conditions can likely initiate the fission process. The artificial fission can be accomplished by a rubber band constricting the body or by simple cutting (Abdel Razek et al., 2007). However, not all sea cucumber species can restore their body parts after artificial fission as compared with natural fission (Dolmatov, 2014).

References

Abdel Razek, F.A., Abdel Rahman, S.H., Mona, M.H., El-Gamal, M.M., Moussa, R.M., 2007. An observation on the effect of environmental conditions on induced fission of the Mediterranean sand sea cucumber, *Holothuria arenicola* in Egypt. Beche-de-mer Inf. Bull. 26, 33–34.

Ahmed, M.I., 2009. Morphological, Ecological and Molecular Examination of the Seacucumber Species along the Red Sea Coast of Egypt and Gulf of Aqaba: With the Investigation of the Possibility of Using DNA Barcoding Technique as a Standard Method for Seacucumber ID. Doctoral dissertation. The University of Hull.

Baccetti, B., Rosati, F., 1968. The fine structure of the polian vesicles of holothurians. Z. für Zellforsch. Mikrosk. Anat. 90, 148–160.

Brown, W.I., Shick, J.M., 1979. Bimodal gas exchange and the regulation of oxygen uptake in holothurians. Biol. Bull. 156 (3), 272–288.

Byrne, M., 2001. The Morphology of Autotomy Structures in the Sea Cucumber Eupentacta Quinquesemita before and During Evisceration, 849–863 pp.

Carlson, B.M. (Ed.), 2011. Principles of Regenerative Biology. Elsevier.

Díaz-Balzac, C.A., Abreu-Arbelo, J.E., García-Arrarás, J.E., 2010. Neuroanatomy of the tube feet and tentacles in *Holothuria glaberrima* (Holothuroidea, Echinodermata). Zoomorphology 129, 33–43.

Díaz-Balzac, C.A., Lázaro-Peña, M.I., Vázquez-Figueroa, L.D., Díaz-Balzac, R.J., García-Arrarás, J.E., 2016. Holothurian nervous system diversity revealed by neuroanatomical analysis. PloS One 11.

Dolmatov, I.Y., 2014. Asexual reproduction in holothurians. Sci. World J. 2014.

Dolmatov, I.Y., Eliseikina, M.G., Bulgakov, A.A., Ginanova, T.T., Lamash, N.E., Korchagin, V.P., 1996. Muscle regeneration in the holothurian *Stichopus japonicus*. Roux's Arch. Dev. Biol. 205, 486–493.

Eaves, A.A., Richard Palmer, A., 2003. Widespread cloning in echinoderm larvae. Nature 425, 146.

Eliseikina, M.G., Magarlamov, T.Y., 2002. Coelomocyte morphology in the holothurians *Apostichopus japonicus* (Aspidochirota: Stichopodidae) and *Cucumaria japonica* (Dendrochirota: Cucumariidae). Russ. J. Mar. Biol. 28, 197–202.

Ezhova, O.V., Ershova, N.A., Malakhov, V.V., 2017. Microscopic anatomy of the axial complex and associated structures in the sea cucumber *Chiridota laevis* Fabricius, 1780 (Echinodermata, Holothuroidea). Zoomorphology 136, 205–217.

Feral, J.P., Massin, C., 1982. Digestive Systems: Holothuroidea: Echinoderm Nutrition, pp. 191–212.

Guerrero Guerrero, A., Rodríguez Forero, A., 2018. Histological characterization of skin and radial bodies of two species of genus *Isostichopus* (Echinodermata: Holothuroidea). Egypt J. Aquat. Res. 44, 155–161.

Gaudron, S., 2006. Observation of natural spawning of *Bohadschia vitiensis*. Beche-de-mer Inf. Bull. 24, 54.

Hamel, J.F., Mercier, A., 1996. Early development, settlement, growth, and spatial distribution of the sea cucumber *Cucumaria frondosa* (Echinodermata: Holothuroidea). Can. J. Fish. Aquat. 53, 253–271.

Hetzel, H.R., 1963. Studies on holothurian coelomocytes. I. A survey of coelomocyte types. Biol. Bull. 125 (2), 289–301.

Hyman, L.H., 1955. Echinodermata. The Coelomate Bilateria, 763 pp.

Inoue, M., Birenheide, R., Koizumi, O., Kobayakawa, Y., Muneoka, Y., Motokawa, T., 1999. Localization of the neuropeptide NCIWYamide in the holothurian nervous system and its effects on muscular contraction. Proc. Biol. Sci. 266, 993–1000.

Iwalaye, O.A., Moodley, G.K., Robertson-Andersson, D.V., 2020. The possible routes of microplastics uptake in sea cucumber *Holothuria cinerascens* (Brandt, 1835). Environ. Pollut. 264, 114644.

Jangoux, M., Lawrence, J.M., 1982. Echinoderm Nutrition. CRC Press.

Kamarudin, K.R., Rehan, M.M., 2015. Morphological and molecular identification of holothuria (*Merthensiothuria*) *leucospilota* and *Stichopus horrens* from Pangkor Island, Malaysia. Trop. Life Sci. Res. 26, 87–99.

Kamenev, Y.O., Dolmatov, I.Y., Frolova, L.T., Khang, N.A., 2013. The morphology of the digestive tract and respiratory organs of the holothurian *Cladolabes schmeltzii* (Holothuroidea, Dendrochirotida). Tissue Cell 45, 126–139.

Lambert, P., 1997. Sea Cucumbers of British Columbia, Southeast Alaska and Puget Sound. UBC Press.

Mashanov, V.S., Frolova, L.T., Dolmatov, I.Y., 2004. Structure of the digestive tube in the holothurian *Eupentacta fraudatrix* (Holothuroidea: Dendrochirota). Russ. J. Mar. Biol. 30, 314–322.

Mashanov, V.S., Zueva, O.R., Heinzeller, T., Dolmatov, I.Y., 2006. Ultrastructure of the circumoral nerve ring and the radial nerve cords in holothurians (Echinodermata). Zoomorphology 125, 27–38.

Massin, C., Lane, D., 1991. Description of a new species of sea cucumber (Stichopodidae: Holothuridae) from the Eastern Indo-Malayan Archipelago: *Thelenota rubralineata* n. sp. Micronesica 24, 57–64.

Motokawa, T., 1984. Connective tissue catch in echinoderms. Biol. Rev. 59 (2), 255—270.

Newell, R.C., Courtney, W.A.M., 1965. Respiratory movements in *Holothuria forskali* Delle Chiaje. J. Exp. Biol. 42, 45—57.

Purcell, S., Samyn, Y., Conand, C., 2012. Commercially Important Sea Cucumbers of the World, 223 pp.

Ramírez-Gómez, F., Aponte-Rivera, F., Méndez-Castaner, L., García-Arrarás, J.E., 2010. Changes in holothurian coelomocyte populations following immune stimulation with different molecular patterns. Fish Shellfish Immunol. 29, 175—185.

Reich, M., 2015. Different pathways in early evolution of the holothurian calcareous ring? Prog. Echinoderm Palaeobiol. 137—145.

Rosati, F., 1970. The fine structure of the alimentary canal of holothurians: 2. the uptake of ferritin and iron-dextran. Monit. Zool. Ital. - Ital. J. Zool. 4, 107—113.

Ruppert, E.E., Barnes, R.D., 1994. Invertebrate Zoology: A Functional Approach. Saunders College Publishing., New York, 950—966 pp.

Truong, T., Le, T., 2019. Characterization of six types of dried sea cucumber product from different countries. Int. J. Food Sci. Agric. 3, 220—231. https://doi.org/10.26855/ijfsa.2019.09.011.

Uthicke, S., Conand, C., 2005. Amplified fragment length polymorphism (AFLP) analysis indicates the importance of both asexual and sexual reproduction in the fissiparous holothurian *Stichopus chloronotus* (Aspidochirotida) in the Indian and Pacific Ocean. Coral Reefs 24, 103—111.

Vergara, W., Rodríguez, A., 2015. Histology of the digestive tract of three species of sea cucumber *Isostichopus badionotus*, *Stichopus* sp. and *Stichopus hermanni* (Aspidochirotida: Stichopodidae). Rev. Biol. Trop. 63 (4), 1021—1033.

Woo, S.P., Yasin, Z., Tan, S.H., Kajihara, H., Fujita, T., 2015. Sea cucumbers of the genus Stichopus Brandt, 1835 (Holothuroidea, Stichopodidae) in straits of malacca with description of a new species. ZooKeys 2015, 1—26.

Zhang, X., Sun, L., Yuan, J., Sun, Y., Gao, Y., Zhang, L., Li, S., Dai, H., Hamel, J.F., Liu, C., Yu, Y., 2017. The sea cucumber genome provides insights into morphological evolution and visceral regeneration. PLoS Biol. 15 (10), e2003790.

Behaviour and ecology

3

3.1 Evisceration

Evisceration is a form of autotomy that has been observed in sea cucumbers. This behaviour involves casting off the internal organs once they feel stress or as a defensive mechanism (Figs. 3.1 and 3.2) (Byrne, 1985; Dawbin, 1948; Dolmatov et al., 2012; Flammang et al., 2002). The sea cucumber *Holothuria difficilis*, for instance, can eviscerate its Cuverian tubules as toxins accurately towards predators (Bakus, 1968). The internal viscera are considered a toxin that can lead to mortality in fish

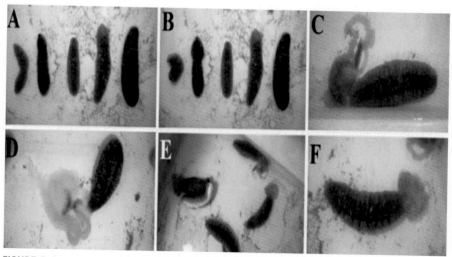

FIGURE 3.1

The evisceration of sandfish *Holothuria scabra* in styrofoam boxes and after transportation by styrofoam box. (A and B): Sandfish in styrofoam boxes. (C–F): Sandfish evisceration in styrofoam boxes.

Credit: Tuwo, A., Yasir, I., Tresnati, J., Aprianto, R., Yanti, A., Bestari, A.D., Syafiuddin, Nakajima, M., 2019.
Evisceration rate of sandfish Holothuria scabra during transportation, in: IOP Conference Series: Earth and Environmental Science. pp. 12039. https://doi.org/10.1088/1755-1315/370/1/012039. Licensed under CC BY 3.0.

FIGURE 3.2

The sea cucumber *Holothuria forskali* emitting Cuvierian tubules.

Credit: Rpillon, CC BY-SA 3.0 via Wikimedia commons.

(Bakus, 1968). However, evisceration is not always successful in avoiding predation, presumably because of the lack of a refuge (Byrne, 1985). Additionally, some sea cucumbers were observed to eviscerate seasonally in nature, which may be ascribed to environmental drives (Swan, 1961).

Evisceration differs between sea cucumber species. Sea cucumbers from the order Dendrochirotida expel their internal organs through their mouth, mainly the intestines, haemal vessels, tentacular crown and sometimes the gonads. However, sea cucumbers from the order Aspidochirotida eviscerate posteriorly through their anus, mainly the respiratory tree, the intestines and the haemal system (Byrne, 1985). The evisceration can be induced artificially by the injection of a distilled water or KCl into the body cavity (Dawbin, 1948; García-Arrarás et al., 2018). The body wall and the eviscerated organs become soft, preparing to extrude the viscera (Motokawa, 1984; Byrne, 1985).

3.2 Regeneration

The regeneration ability of sea cucumbers has attracted the attention of biologists for decades. Regeneration occurs in holothurians in the adults and the larvae stage as well. Sea cucumbers can repair and regenerate their organs (intestines, gonads and respiratory tree) or even reproduce asexually to a new adult (Carnevali, 2006; Dolmatov et al., 2012; García-arrarás and Greenberg, 2001). Regeneration ability differs between holothurians; for instance, the sea cucumber *Sclerodactyla briareus* can regenerate the nerve ring after a few weeks (García-Arrarás et al., 2018). Also, the sea cucumber *Leptosynapta crassipatina* can grow to a complete animal from the disk that surrounds the mouth containing the calcareous ring and the nerve ring (Smith, 1971). Furthermore, the sea cucumber *Apostichopus japonicus* is capable of regenerating only the posterior part, whereas *Holothuria atra* is capable of regenerating both posterior and anterior parts (Dolmatov et al., 2012). Moreover, the regeneration rate of the gut varies between species. The sea cucumber *Holothuria scabra* can feed after 7 days of evisceration (Bai, 1971), whereas feeding was observed in *Australostichopus mollis* after 145 days (Dawbin, 1948). This variation is likely because of the genetic divergence (Dolmatov et al., 2012).

After evisceration, sea cucumbers start regenerating their organs; the digestive tract and the organs associated with it are the first organs that start regeneration (García-Arrarás et al., 2018; García-arrarás and Greenberg, 2001). The distal edge of the mesenteries, which remain intact and attached to the body wall after eviscerating the digestive tract, plays a vital role in the regeneration of the intestines in holothurians (Figs. 3.3 and 3.4) (Bai, 1971; Byrne, 1985; Dawbin, 1948; García-Arrarás et al., 1998; Kille, 1935; Leibson, 1992; Mosher, 1956; Smith, 1971). Specifically, a solid cord develops and grows along the mesenterial edge from the oesophagus to the cloaca. Then, the intestinal lumen is formed, and the epithelial cells become recognisable and begin to form a full lining (García-arrarás and Greenberg, 2001; Okada and Kondo, 2019). After that, the haemal vessels develop within the gut and extend to the oesophagus (Figs. 3.3 and 3.4) (Bai, 1971; Dawbin, 1948).

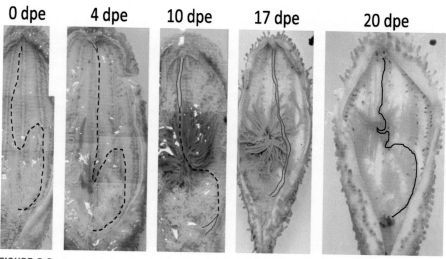

FIGURE 3.3

Internal morphology of regenerating animals. Black broken line: the edge of mesentery; solid red line: area along the edge of the mesentery, with thickened gut rudiments; black line: grown gut rudiment. *dpe*, days post evisceration.

Credit: Okada, A., Kondo, M., 2019. Regeneration of the digestive tract of an anterior-eviscerating sea cucumber, Eupentacta quinquesemita, *and the involvement of mesenchymal-epithelial transition in digestive tube formation. Zool. Lett. 5, 21. https://doi.org/10.1186/s40851-019-0133-3.*

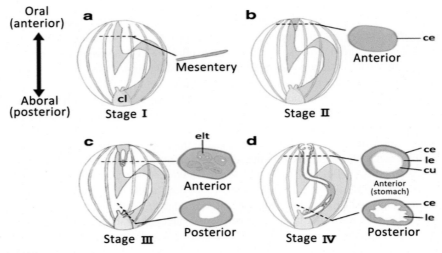

FIGURE 3.4

Schematic diagram of regeneration of the digestive tube in the sea cucumber *Eupentacta quinquesemita*. (A) Stage I (0 dpe): just after evisceration. Only the mesentery (grey shaded area) and the cloaca (cl; yellow) remain in the body cavity. (B) Stage II (1—4 dpe): a regenerating tissue (gut rudiment) appears at the anterior side as a mass of mainly mesenchymal cells surrounded by coelomic epithelium. (C) Stage III (5—14 dpe): multiple cavities are formed in the anterior regenerating tissue, and these coalesce with each other to form lumens. (D) Stage IV (14—20 dpe): when regeneration is more progressed, the gut rudiment (thickened tissue on the mesentery) becomes continuous between the anterior and posterior sides. The figure shows that the anterior and posterior lumens are not yet connected, but later a single continuous tube is completed. Intestines differentiate according to their position in the digestive tract. In the stomach, a muscle layer (not shown) develops and luminal epithelium (le) is covered with cuticles (cu), as in intact tissues. *ce*, coelomic epithelium; *cl*, cloaca; *cu*, cuticles; *dpe*, days post evisceration; *elt*, epithelium-like tissue; *le*, luminal epithelium.

Credit: Okada, A., Kondo, M., 2019. Regeneration of the digestive tract of an anterior-eviscerating sea cu-cumber, Eupentacta quinquesemita, *and the involvement of mesenchymal-epithelial transition in digestive tube formation. Zool. Lett. 5, 21. https://doi.org/10.1186/s40851-019-0133-3.*

The regeneration process of the intestines includes wound healing, blastema formation, lumen formation, intestine differentiation and intestine growth (Zhang et al., 2017). For the respiratory tree, it is formed from the anterior part of the cloaca and the internal lumen forms tubes that branch into smaller ampullae (Bai, 1971; Dawbin, 1948). For the gonads, the sea cucumbers from the order Aspidochirota eviscerate through the cloaca, and the gonadal base is retained, which forms the gonad later. However, the sea cucumbers from the order Dendrochirotida eviscerate from their mouth, and the gonadal basis remains attached to the dorsal mesenteries,

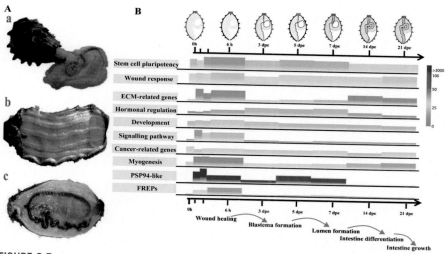

FIGURE 3.5

(A) Regeneration diagram showing *Apostichopus japonicus* (A) undergoing evisceration, (B) immediately post evisceration and (C) after complete recovery. (B) A heatmap showing the expression profile of molecular events applicable to the intestinal regeneration of *A. japonicus*. High-expressed and low-expressed genes are labelled in red and green colours. *dpe*, days post evisceration; *ECM*, extracellular matrix; *FREPs*, fibrinogen-related proteins; *PSP94*, prostatic secretory protein-94.

Credit: Zhang, X., Sun, L., Yuan, J., Sun, Y., Gao, Y., Zhang, L., Li, S., Dai, H., Hamel, J.F., Liu, C., Yu, Y., 2017. The sea cucumber genome provides insights into morphological evolution and visceral regeneration. PLoS Biol. 15, e2003790. https://doi.org/10.1371/journal.pbio.2003790.

which will form the new gonads (Dawbin, 1949; Mosher, 1956; Leibson, 1992; Garcı'a-Arrara's et al., 1998; García-Arrarás and Greenberg, 2001).

Some genes were found in high expression during the regeneration process in the sea cucumber *A. japonicus*. Zhang et al. (2017) reported that signalling pathways (Wnt, bone morphogenetic protein-related genes and epidermal growth factor-related genes), extracellular matrix (ECM)-related genes (collagen and fibropellin) and myogenesis-related genes (tubulin) are activated at the early stage of regeneration [0—3 days post evisceration (dpe)]. During the middle stage of regeneration (3—7 dpe), factors related to hormonal regulation are up-regulated. In the late stages of regeneration, ECM-related genes (tenascin, FRAS1 and collagen) and myogenesis-related genes (actin and myosin) show up-regulated expression (Fig. 3.5) (Zhang et al., 2017). Also, Mashanov et al. (2010) concluded that the expression of pro-cancer genes, such as survivin and mortalin, is involved in the regeneration process, which coincides in time with drastic de-differentiation and a burst in cell division and apoptosis (Fig. 3.6) (Mashanov et al., 2010).

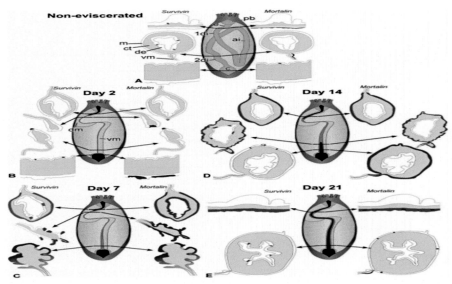

FIGURE 3.6

The anatomic features of the non-eviscerated (normal) and regenerating digestive tube and the expression patterns of survivin and mortalin in the sea cucumber *Holothuria glaberrima*. (A) Non-eviscerated animals. (B)–(E) Regenerating animals on days 2, 7, 14 and 21, respectively. *1di*, first descending intestine; *2di*, second descending intestine; *ai*, ascending intestine; *c*, cloaca; *ct*, connective tissue layer; *de*, digestive (luminal) epithelium; *dm*, dorsal mesentery; *e*, oesophagus; *m*, mesothelium; *pb*, pharyngeal bulb; *vm*, ventral mesentery. The blue colour indicates in situ hybridisation signal; red colour indicates regenerated tissues; the black colour indicates lumen of the digestive tube.

Credit: Mashanov, V.S., Zueva, O.R., Rojas-Catagena, C., Garcia-Arraras, J.E., 2010. Visceral regeneration in a sea cucumber involves extensive expression of survivin and mortalin homologs in the mesothelium. BMC Dev. Biol. 10. https://doi.org/10.1186/1471-213X-10-117.

3.3 Aestivation

Aestivation is a dormant state, used as a survival strategy when facing very hot or arid conditions. Aestivation has been observed in many vertebrates and invertebrates, for instance, the African lungfish (Delaney et al., 1974), some species of hylid frog, spadefoot toads, some species of land snails (Carvalho et al., 2010) and sea cucumbers (Liu et al., 1996). Sea cucumbers aestivate when the water temperature becomes high and hibernate when the water temperature becomes low. Günay et al. (2015) reported that the sea cucumber *H. tubulosa* entered the aestivation status under seawater temperature of 30°C and entered the hibernation status under seawater temperature of 15°C. Under the hibernation and aestivation status, little or no growth was observed at 15°C and 30°C (Günay et al., 2015).

During aestivation, sea cucumbers seek refuges, stiffen their body walls, stop feeding gradually and undergo gastrointestinal degradation, lose weight, change immunity, reduce oxygen consumption and redistribute their energy during this phase until the surrounding condition improves (Motokawa, 1984; Chen et al., 2016). Remarkable changes in the physiology of the sea cucumber occur during aestivation. For instance, in the sea cucumber *A. japonicus*, the intestine undergoes deterioration and becomes less than half of its original length. Also, a reduction in energy consumption and storage occurs (Gao et al., 2009). As a result, the energy budget, oxygen consumption and the ammonia execration rate are decreased (Yuan et al., 2007).

Although silencing of the gene expression occurs, the regulation of few genes is maintained or increased. Most of the genes are differentially expressed in the body wall, followed by muscles, respiratory tree and the intestines (Li et al., 2018). Li et al. (2018) suggested that the aestivation of the sea cucumber is a dormancy status that might be controlled by a clock gene-controlled process possibly triggered by early growth response protein 1 (Egr1) and Kruppel-like factor 2 (Klf2) (Li et al., 2018). Klf2 and Egr1 are the most significant transcription factors that are expressed mainly in the body wall during aestivation. Egr1 can regulate the clock gene cryptochrome circadian regulator 1 (Cry1), which is the primary repressor gene that can propel the animal into the sleep phase (Li et al., 2018). Cry1 is highly expressed during early and deep aestivation, and its expression decreased after aestivation. The expression of primary activators Clock and Bmal1 is suppressed during deep aestivation. It is reactivated with emerging arousal from aestivation, which is associated with the activity of Cry1 (Fig. 3.7) (Li et al., 2018). Furthermore, the expression of the DNMT1 gene and DNA methyltransferase 1 increases in the intestines during aestivation. Also, the expression of the MBD2/3 gene and methyl-CpG-binding domain gene type 2/3 increases in the tissues of the respiratory tree during aestivation (Zhao et al., 2013).

3.4 Population genetics

Genetic diversity plays a vital role in the survival and adaptability of species. It can arise through the process of random mutation or natural selection. A new mutation will increase genetic diversity in the short term, and natural selection for, or against, a trait can occur with changing the environment. However, gene flow may occur through the dispersal of gametes and individuals with subsequent successful reproduction and thus oppose local differentiation through introducing novel alleles to the local population (Arndt and Smith, 1998; Wright, 2005; Nevo, 2001) (Fig. 3.8). Understanding the genetic structure of the sea cucumber populations is essential for preserving the genetic diversity in natural populations when restocking and managing the fisheries of sea cucumber species (Purcell, 2004).

It is assumed that animals with a long larvae period have high dispersal capability and high gene flow among populations. Barriers to the dispersal can be

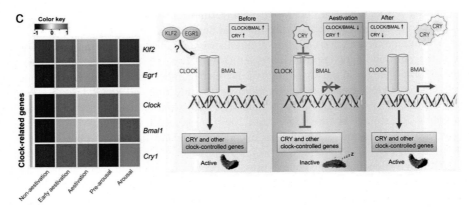

FIGURE 3.7

Expression heatmap of Kruppel-like factor 2 (Klf2), early growth response protein 1 (Egr1) and clock-related genes during different aestivation states (according to quantitative polymerase chain reaction results) and the inferred clock gene-controlled regulation model. Klf2 and Egr1 may trigger the upregulation of cryptochrome circadian regulator 1 (Cry1) (either directly or indirectly through Clock and Bmal1) during sea cucumber aestivation, which propels the animal into an extended sleep phase, and decreased Cry1 expression makes the animal awaken from aestivation.

Credit: Li, Y., Wang, R., Xun, X., Wang, J., Bao, L., Thimmappa, R., Ding, J., Jiang, J., Zhang, L., Li, T., Lv, J., 2018. Sea cucumber genome provides insights into saponin biosynthesis and aestivation regulation. Cell Discov. 4. https://doi.org/10.1038/s41421-018-0030-5.

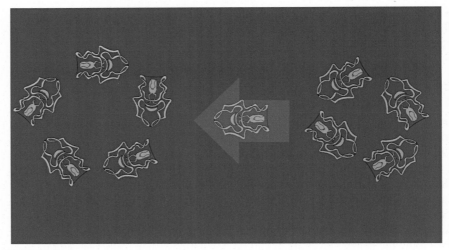

FIGURE 3.8

Gene flow is the transfer of alleles from one population to another through the dispersal of gametes or larvae with subsequent successful reproduction.

Credit: Mohamed Mohsen.

biological, such as behavioural patterns; physical, such as temperature or currents (Lohr, 2003) or biographic barriers arising from sea-level changes (Mirams et al., 2011). The population of two sea cucumber species (i.e. *Cucumaria miniata* and *Cucumaria pseudocurata*) was examined in the northeastern Pacific between Alaska and California based on mitochondrial DNA sequence analysis. The authors demonstrated that changes in the life history patterns caused a difference in the genetic structure of populations. Given that the larvae of the sea cucumber *C. miniata* is pelagic larvae, which spend less than 2 weeks in the water column, its population exhibits an unstructured population. In contrast, the absence of the pelagic larvae stage in the sea cucumber *C. pseudocurata* results in a highly structured population (Arndt and Smith, 1998). Also, the sea cucumber *Isostichopus focus* is a sea cucumber with a pelagic larvae period. Its population was examined from the Galápagos archipelago, mainland Ecuador and the Gulf of California, Mexico, to determine the level of gene flow among these populations. The author demonstrated that there are genetic similarities among southern populations and high inferred levels of gene flow, suggesting larval dispersal between mainland and island. Gene flow was found between Galapagos and Ecuador, which appeared to be connected genetically (Lohr, 2003).

The estimation of genetic diversity can aid the implementation of efficient conservation programmes through preserving genetic diversity when releasing the hatchery-produced larvae (Uthicke and Purcell, 2004; Chen et al., 2008). For instance, the genetic heterogeneity of the sea cucumber *H. scabra* was examined in nine sites in New Caledonia for consideration of releasing hatchery-produced larvae that were collected at one of these sites. Depending on the genetic differentiation between stocks of *H. scabra*, the authors concluded that hatchery-produced juveniles that arise from six sites could be released into any of these sites without affecting the genetic diversity of the species in that area (Uthicke and Purcell, 2004).

Furthermore, the estimation of genetic diversity can introduce useful information of the natural population structure and the dispersal of sea cucumbers over a variety of geographic scales and thus, aid in the selective breeding, conservation and management of sea cucumber resources (González-Wangüemert et al., 2015; Kim et al., 2008; Ochoa-Chávez et al., 2018; So et al., 2011; Soliman et al., 2012; Yao et al., 2007). The genetic variation of 15 populations of the sea cucumber *Holothuria nobilis* was examined in the Great Barrier Reef. The authors concluded that there is high gene flow between populations and that sexually produced larvae are the only source of *H. nobilis* population over the tested locations, which is subjected to details in their hydrographic connection (Uthicke and Benzie, 2000). Also, the genetic structure of the sea cucumber *Holothuria poli* population was examined across the Mar Menor coastal lagoon and nearby marine areas. However, with the unstable environment in the lagoon, *H. poli* showed high values of genetic diversity in this region besides the presence of exclusive haplotypes. The authors concluded that the population genetic variability promoted by gene flow could increase the sea cucumber potential to adapt to the changing conditions of this lagoon environment (Vergara-Chen et al., 2010).

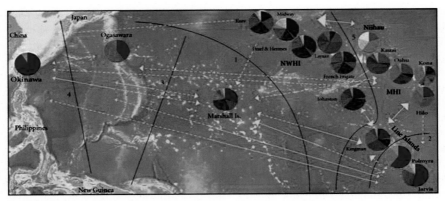

FIGURE 3.9

Map of the northern Central and West Pacific. Pie charts represent haplotype frequencies overlaid on sites. Black lines are major gene flow constraints drawn by BARRIER, numbered from strongest to weakest. Yellow arrows represent directional migration rates estimated by MIGRATE. Effective migration rate estimates (N_eM) with modes between 0.01 and 0.49 are represented by small arrows, rates between 0.5 and 0.99 are represented by medium arrows and rates over 1.0 are represented by large arrows. Solid lines represent migration posterior probability distributions, where the 50% credibility set does not include zero. Dashed lines represent migration posterior probability distributions, where the 50% credibility set includes zero, but the distribution peak is greater than zero.

Credit: Skillings, D.J., Bird, C.E., Toonen, R.J., 2011. Gateways to Hawai 'i: genetic population structure of the tropical sea cucumber Holothuria atra. J. Mar. Biol. 2011, 1–16. https://doi.org/10.1155/2011/783030.

Moreover, the population of *H. atra* was assessed for determining the appropriate scale for coral reef management. The sea cucumber was sampled from the Hawaiian Islands, Line Islands, Marshal Islands, Bonin Islands and the Ryukyu Islands. The authors concluded that *H. atra* has a hierarchical, fine-scale population structure driven primarily by between-archipelago barriers but with significant differences between sites and within an archipelago as well. Hawaiian Archipelago and Kingman Reef are ancestral populations in the region, with migration moving out of these periphery archipelagos towards a less diverse Central Pacific rather than the reverse. Finally, Kingman Reef is the most likely stepping stone between Hawaii and the rest of the Pacific for *H. atra* (Fig. 3.9) (Skillings et al., 2011).

Population genetics analysis was also employed to understand the genetic structure and diversity in the colour variation of the sea cucumber *A. japonicus*, given the red colour is the most valuable. The genetic differentiation among three colour variants (red, green and black) of the sea cucumber *A. japonicus* was explored from four localities in Japan. The red type inhabits the gravel bed offshore, whereas the green and black types inhabit the sand-muddy bottom inshore. The study concluded that the red type has a definite genetic differentiation from the other colour types. The authors explained that this genetic independence is likely because of the differences in microhabitats and other reproductive isolation systems, which prevent free mating between the three colour morphs (Kan-No and Kijima, 2003). Also, the genetic structure of wild and hatchery-produced green type *A. japonicus* was examined in Korea and China, and wild red type from Korea using nine microsatellite markers. The authors concluded that the red and green variants are reproductively isolated (Kang et al., 2011).

3.5 Sea cucumbers interaction with the surrounding environment

3.5.1 Bioturbation

Bioturbation is defined as mixing or reworking the sediment biogenically (Kristensen et al., 2012). Sea cucumbers disturb the upper layer of the sediment through ingestion and egestion, locomotion, respiration and burrowing (Fig. 3.10) (Uthicke, 1999). The biogenic mixing level of sea cucumbers varies between species. Some sea cucumbers tend to bury themselves totally or partially in sediment, which leads to mixing the upper centimetres of the sediment. Thus, they have a significant role in bioturbation, and their overexploitation presents a threat to the bottom fauna (Dar, 2004; Hanafy, 2011). For instance, the sea cucumber *Holothuria arenicola* mixes about 10 cm of the upper layers of sediment (Hammond, 1982; Pawson and Caycedo, 1980). Also, the sea cucumbers *H. atra* and *Stichopus chloronotus* process the total upper 5 mm of coral reef sediments per annum (Uthicke, 1994), whereas the sea cucumber *Holothuria whitmaei* processes only 14.1% of the available sediment, on average (Shiell and Knott, 2010).

Mixing the sediment helps in releasing nutrients, biodiversity driving and decrease of organic matter (Adámek and Maršálek, 2013). Sea cucumbers ingest sediment and organic matter and grind it into smaller particles. This behaviour assists with breaking up and shuffling the top layer of sediment (Sloan and von

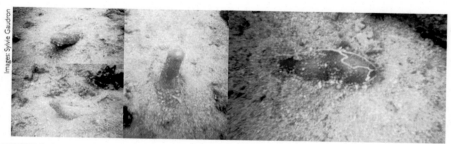

FIGURE 3.10

Bioturbation activity during spawning of the sea cucumber *Bohadschia vitiensis*.

Credit: Sylvie Guardon.

Bodungen, 1980). Also, it facilitates the recycling of nutrients at the sediment-water interface, thereby contributing to the overall health of habitats (Uthicke, 2001). Hou et al. (2017) reported that the sea cucumber bioturbation modifies the sediment conditions through the digestive processes and mechanical disturbance, which leads to changes in inorganic carbon dissolution.

3.5.2 Benthic oxygen regeneration

Benthic macroinvertebrates support oxygen flux to the sediment and the benthic fauna through feeding, movement and respiration (Adámek and Maršálek, 2013; Pischedda et al., 2008). Oxygen is required for the respiration of benthic fauna and the decomposition of organic matter. Increasing the availability of oxygen through bioturbation can accelerate organic matter degradation. Investigation showed that the old substances exposed by bioturbation decompose faster by 3.6 times with oxygen than without oxygen (Hulthe et al., 1998). Sea cucumbers alter the physical characteristic of the sediments and increase the surface area of the sediment, which helps to drive water flow into the sediment, delivering O_2 and degradable materials and preventing the formation of anaerobic conditions (Işgören-Emiroğ;lu and Günay, 2007; Lee et al., 2017).

3.5.3 Organic matter decomposition

Organic matter refers to the carbon-based compounds that result from the remains of organisms or their wastes. Increasing organic matter in the water can deplete the oxygen and cause eutrophication problems (Mackin and Swider, 1989). Sea cucumbers have a significant role in decreasing organic matter in the sediments; therefore, they help to recycle and regenerate nutrients.

How can sea cucumbers decrease organic matter? One possible scenario is that sea cucumbers initiate the growth of the microbial decomposers, which, in turn, help to decompose the organic matter (MacTavish et al., 2012). The other scenario is that some holothurians select detritus and organic-rich sediment as the organic matter in their intestines were found to be higher than of the surrounding sediments (Hauksson, 1979; Moriarty, 1982; Slater et al., 2011; Zamora and Jeffs, 2011; Zhao and Yang, 2010). Although some sea cucumbers were found to be non-selective feeders (Hammond, 1982; Uthicke and Karez, 1999), they defecate sediment lower organic matter than they fed (Mercier et al., 1999; Purcell et al., 2016). Therefore, high rates of organic matter in the sediments can be an option to increase the density of sea cucumbers (Zamora and Jeffs, 2011). The sea cucumber *Stichopus tremulus* is reported to consume about 0.6 dry sediment/kg year (Hauksson, 1979). Also, *A. mollis* is reported to consume about 3/kg sediment year (Slater and Carton, 2009). Thus, sea cucumbers play an important role in sediment bioremediation.

3.5.4 Algal bloom cleaning

An algal bloom is defined as the rapid increase of the algae population in the water, which results in discolouring the water to their pigments. Whether they are nontoxic or noxious, they cause oxygen depletion once decaying, or they can clog the gills of the filter-feeding animals in some cases. Moreover, they compete for available oxygen, which represents a problem, especially in aquaculture management (Shumway, 1990). Sea cucumbers help to decrease algal bloom directly through feeding on the algae or indirectly through decreasing the organic load. The filter-feeding sea cucumbers can decrease the organic load in the water column through suspension feeding (Nelson et al., 2012). For deposit feeder sea cucumbers, Michio et al. (2003) found a decrease in organic matter and algal bloom with the occurrence of sea cucumber *A. japonicus*. Without the sea cucumbers, the algae had covered the bottom after 2 weeks. Two possible scenarios are proposed for decreasing algal bloom by deposit feeder sea cucumbers. The first scenario is the grazing by sea cucumber in which the sea cucumber can feed on the upper layer of the sediment (Slater and Carton, 2009). The second scenario is the uptake of suspended particles through the mouth (Mohsen et al., 2020). The deposit feeder sea cucumber (*A. japonicus*) could also uptake particles from the water column, which might have a significant role in decreasing the organic content of the water (Mohsen et al., 2020). Consistently, the 18s rDNA analysis revealed that sea cucumber *A. japonicus* uptake more nutrients from the water column rather than from the sediment (Zhang et al., 2016).

3.5.5 Water chemistry improving

Ocean acidification happens via the transference of atmospheric CO_2 to the ocean, creating chemical reactions that lead to a decrease in the pH. Ocean acidification impedes calcification, which is considered a threat to marine ecosystems and potentially for coral reefs (Schmutter et al., 2017). Sea cucumbers puffer the effect of acidification throughout calcium carbonate ($CaCO_3$) dissolution in their gut and ammonia (NH_3) excretion (Schneider et al., 2015; Wolfe et al., 2018). Sediment dissolution and respiration production by *Stichopus herrmanni* buffers ocean acidification (Wolfe et al., 2018). Schneider et al. (2015) reported that sea cucumbers *S. herrmanni* and *Holothuria leucospilota* cause alkalinity increasing by 97 and 47 µmol/kg, respectively, which is a result of $CaCO_3$ dissolution (81 µmol/kg) and NH_3 secretion (34 µmol/kg). NH_3 is secreted as a digestion by-product that uptakes a proton (H+) and ionises to NH_4, which increases the alkalinity. These nutrients buffer the effect of ocean acidification and help the reef to classify by accumulating $CaCO_3$ more than or equal to the eroding rates (Schneider et al., 2015).

3.5.6 Mariculture waste bioremediation

Increasing aquaculture production led to an increase in the faeces of the cultured animals and the excess feed in the habitats, which may cause severe aquatic pollution (Cao et al., 2007). The excess feed and faeces are viewed as a challenge to manage in the farms, which leads to an increase in the nutrients load in the marine environment, consuming the oxygen, increasing the ammonium and inhabiting denitrification (Cao et al., 2007; Ziemann et al., 1992). One possible solution is using the integrated culture with sea cucumbers (Inui et al., 1991). Using the integrated aquaculture system is considered as a promising solution that can reduce the nutrients load in the aquatic environment and increase the profit as well. Sea cucumbers are highlighted as a promising option in the integrated systems that can decrease feed expenses, increase the production per unit area and decrease organic wastes load (Zamora et al., 2016).

Previous research concluded that rearing the sea cucumber *Parastichopus californicus* inside the net of salmon helped in cleaning a wide area of the net, which assimilated organic matter and amino acids more efficiently than from natural sediment (Ahlgren, 1998). Similarly, Zhou et al. (2006) observed a reduction in the organic content of collected wastes in scallop lantern nets following *A. japonicus* feeding. Also, the presence of sea cucumber *A. mollis* with sediment impacted with bivalve wastes resulted in decreasing the chlorophyll a/phaeopigment ratio, chlorophyll a, total organic carbon and phaeopigment (Slater and Carton, 2009). Likewise, *H. atra* and *S. chloronotus* significantly reduced chlorophyll a concentrations in sediments inoculated with diatoms or cyanobacterial mats (Uthicke, 1999). Furthermore, Ren et al. (2012) reported that the existence of the sea cucumber *A. japonicus* could effectively reduce the accumulation of nutrients in cultured ponds. The nutrients in the pond increased during the period of aestivation or hibernation of the sea cucumber. Moreover, Kang et al. (2003) reported an increase in nutrient cycling using sea cucumbers in polyculture with tank-farmed abalone. These results revealed that sea cucumbers could benefit the removal of biological pollution (Işgören-Emiroğ;lu and Günay, 2007).

3.5.7 Relationship with benthic organisms

Sea cucumbers influence the productivity of many benthic organisms, which is likely because of their ability of nutrient recycling. These nutrients are utilised by many benthic organisms, such as microalgae, macroalgae and bacteria, which increased throughout utilising the released nutrients by sea cucumbers (MacTavish et al., 2012; Uthicke, 2001). Furthermore, the numbers of benthos in the aquaria with sea cucumbers were more than those without them (Michio et al., 2003). Moreover, the absence of the sea cucumber negatively affects the productivity of seagrass, possibly because of nutrient limitations or light prevention from increasing the organic matter (Wolkenhauer et al., 2010).

Sea cucumbers ingest food enriched in nitrogen and excrete ammonium and a small amount of phosphate close to the sediment. The sea cucumbers *H. atra* and *S. chloronotus* execrate nitrogen ($0.52-5.35$ mg/m^2 day) in the form of ammonium through the respiratory tree and the body wall in addition to phosphate in the form of phosphorus ($0.01-0.47$ mg/m^2 day) (Uthicke, 2001). Furthermore, the sea cucumber *A. mollis* fed on mussel deposited mixed with the sediment produced ammonium with mean flux rates of $11-64$ mmol/m^2 h, mean PO_4^{-3} efflux of 0.2 μmol/m^{2}h and net oxygen production of 586 ± 222 μm/m^2 h (MacTavish et al., 2012). Additionally, a small amount of NO_x flux was observed from the sea cucumber, which may be released by the associated fauna (Uthicke, 2001).

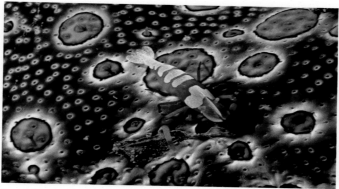

FIGURE 3.11

Emperor shrimp *Periclimenes imperator* on the sea cucumber *Bohadschia ocellata*.

Credit: Nhobgood Nick Hobgood - Own work, CC BY-SA 3.0 via Wikimedia Commons.

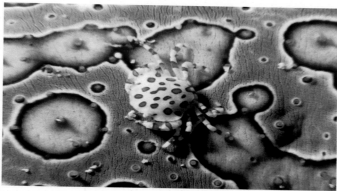

FIGURE 3.12

Swimming crab *Lissocarcinus orbicularis* on the sea cucumber.

Credit: By prilfish CC-BY 2.0 via Wikimedia Commons.

Furthermore, some organisms were found to be associated with sea cucumbers as symbiosis, whether commensal or as parasites (Figs. 3.11—3.14) (Purcell et al., 2016). Commensalism is a long-term biological relationship in which one organism can benefit from the other without harming or benefitting the host (Hartel, 2004). Many commensals species are known to benefit from sea cucumbers, whether endo-commensals or ecto-commensals. Ecto-commensalisms are organisms that live outside on the body of the sea cucumber from several phyla (Eeckhaut et al., 2004; Purcell et al., 2016). Purcell et al. (2016) listed 12 species of pinnotherid crabs that are often associated with several holothuroids. However, endo-commensals benefit from sea cucumbers as a home or a refuge. The pearlfish, for instance,

FIGURE 3.13

Shrimp on a sea cucumber.

Credit: Steve Childs CC-BY 2.0 via Wikimedia Commons.

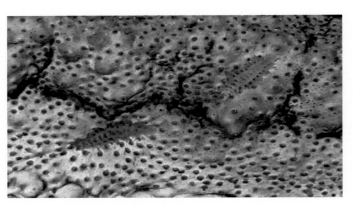

FIGURE 3.14

Polynoid worms on the king sea cucumber.

Credit: Frédéric Ducarme - Own work, CC-BY-SA 4.0 via Wikimedia Commons.

from the family Carapidae consists of species that can live in the coelomic fluid and the respiratory tree of the sea cucumber (Miyazaki et al., 2014).

As parasites, endoparasites from the phylum Platyhelminthes and ectoparasites from the family Eulimidae were found depending on sea cucumbers in their life. Also, the pearlfish fish from the genus *Encheliophis* can feed on the inner tissue of the sea cucumber (Pinn et al., 2014; Purcell et al., 2016). Eeckhaut et al. (2004) reported that about 150 species of metazoans parasitise on sea cucumbers by utilising their bodies as a refuge and food source.

3.5.8 Predators

A wide range of marine predators targets sea cucumbers as prey. Juvenile and larvae are more susceptible to predation than adults. Many species from different taxa can prey on holothurians, such as Gastropoda, Asteroidea, Pisces, Crustacea and Mammalia (Francour, 1997; Purcell et al., 2016). The main predators of sea cucumber species are sea stars, fish and crabs (Francour, 1997; Robinson, 2013). Carnivorous fish can consume a part of the body and reject the rest because of the toxins. Fish can eat the viscera as the sea cucumber eviscerates it to distract the predator (Bakus, 1968). Sea stars are reported to feed on a high number of holothurians (Francour, 1997).

To record the potential of sea star to prey on the sea cucumber, the sea star *Asterina pectinifera* was stocked with a group of the sea cucumber *A. japonicus* in the tanks for 20 days (Fig. 3.15). The sea star sensitively detected and approached the sea cucumber. The sea cucumber showed escaping behaviour without evisceration when touched by the sea star. However, when the sea cucumber was induced to aestivation, the sea star was able to capture the sea cucumber (Fig. 3.16). The sea star digested the body wall of the aestivated sea cucumber in about 17 h (Fig. 3.17). These results suggest that aestivation makes sea cucumber more susceptible to predation by sea stars, which makes sea cucumber easy prey.

FIGURE 3.15

The sea star *Asterina pectinifera* after capturing the aestivated sea cucumber *Apostichopus japonicus*.

Credit: Mohamed Mohsen.

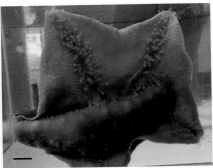

FIGURE 3.16

The sea star digests the body walls of the sea cucumber. Scale bar: 1 cm.

Credit: Mohamed Mohsen.

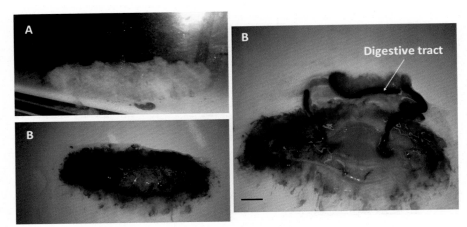

FIGURE 3.17

The sea cucumber after digestion by the sea star. (A) Sea cucumber in the water, (B) Sea cucumber out of the water. Scale bar: 1 cm.

Credit: Mohamed Mohsen.

References

Adámek, Z., Maršálek, B., 2013. Bioturbation of sediments by benthic macroinvertebrates and fish and its implication for pond ecosystems: a review. Aquacult. Int. https://doi.org/10.1007/s10499-012-9527-3.

Ahlgren, M.O., 1998. Consumption and assimilation of salmon net pen fouling debris by the red sea cucumber *Parastichopus californiens*: implications for polyculture. J. World Aquacult. Soc. 29, 133–139. https://doi.org/10.1111/j.1749-7345.1998.tb00972.x.

Arndt, A., Smith, M.J., 1998. Genetic diversity and population structure in two species of sea cucumber: differing patterns according to mode of development. Mol. Ecol. 7, 1053–1064. https://doi.org/10.1046/j.1365-294x.1998.00429.x.

Bai, M.M., 1971. Regeneration in the holothurian, holothuria scabra jager. Indian J. Exp. Biol. 9, 467–471.

Bakus, G.J., 1968. Defensive mechanisms and ecology of some tropical holothurians. Mar. Biol. 2, 23–32. https://doi.org/10.1007/BF00351634.

Byrne, M., 1985. Evisceration behaviour and the seasonal incidence of evisceration in the holothurian *Eupentacta quinquesemita* (Selenka). Ophelia 24, 75–90. https://doi.org/10.1080/00785236.1985.10426621.

Cao, L., Wang, W., Yang, Y., Yang, C., Yuan, Z., Xiong, S., Diana, J., 2007. Environmental impact of aquaculture and countermeasures to aquaculture pollution in China. Environ. Sci. Pollut. Res. https://doi.org/10.1065/espr2007.05.426.

Carnevali, C., 2006. Regeneration in echinoderms: repair, regrowth, cloning. ISJ.

Carvalho, J.E., Navas, C.A., Pereira, I.C., 2010. Energy and water in aestivating amphibians. Prog. Mol. Subcell. Biol. https://doi.org/10.1007/978-3-642-02421-4_7.

Chen, L., Li, Q., Yang, J., 2008. Microsatellite genetic variation in wild and hatchery populations of the sea cucumber (*Apostichopus japonicus* Selenka) from northern China. Aquacult. Res. 39, 1541–1549. https://doi.org/10.1111/j.1365-2109.2008.02027.x.

Chen, M., Li, X., Zhu, A., Storey, K.B., Sun, L., Gao, T., Wang, T., 2016. Understanding mechanism of sea cucumber *Apostichopus japonicus* aestivation: insights from TMT-based proteomic study. Comp. Biochem. Physiol. Genom. Proteonomics 19, 78–89. https://doi.org/10.1016/j.cbd.2016.06.005.

Dar, M.A., 2004. Holothurian Role in the Marine Sediments Reworking Processes. Effects of anthropogenic activities in Red Sea ports on marine environment, Sedimentology of Egypt.

Dawbin, W., 1948. Auto-evisceration and the regeneration of viscera in the holothurian *Sticophus mollis* (Hutton). Trans. R. Soc. N. Z. 77, 497–523.

Dawbin, W.H., 1949. Auto-evisceration and the regeneration of viscera in the holothurian *Stichopus mollis* (Hutton). The Society, pp. 497–523.

Delaney, R.G., Lahiri, S., Fishman, A.P., 1974. Aestivation of the African lungfish *Protopterus aethiopicus*: cardiovascular and respiratory functions. J. Exp. Biol. 61, 111–128.

Dolmatov, I.Y., Khang, N.A., Kamenev, Y.O., 2012. Asexual reproduction, evisceration, and regeneration in holothurians (Holothuroidea) from Nha Trang Bay of the South China Sea. Russ. J. Mar. Biol. 38, 243–252. https://doi.org/10.1134/S1063074012030042.

Eeckhaut, I., Parmentier, E., Becker, P., Gomez da Silva, S., Jangoux, M., 2004. Parasites and biotic diseases in field and cultivated sea cucumbers. In: Advances in Sea Cucumber Aquaculture and Management.

Flammang, P., Ribesse, J., Jangoux, M., 2002. Biomechanics of adhesion in sea cucumber Cuvierian tubules (Echinodermata, Holothuroidea). Integr. Comp. Biol. 1107–1115. https://doi.org/10.1093/icb/42.6.1107.

Francour, P., 1997. Predation on holothurians: a literature review. Invertebr. Biol. 116, 52. https://doi.org/10.2307/3226924.

Gao, F., Yang, H., Xu, Q., Wang, F., Liu, G., 2009. Effect of water temperature on digestive enzyme activity and gut mass in sea cucumber *Apostichopus japonicus* (Selenka), with special reference to aestivation. Chin. J. Oceanol. Limnol. 27, 714–722. https://doi.org/10.1007/s00343-009-9202-3.

García-ArraráS, J.E., Estrada-Rodgers, L., Santiago, R., Torres, I.I., Díaz-Miranda, L., Torres-Avillán, I., 1998. Cellular mechanisms of intestine regeneration in the sea cucumber, Holothuria glaberrima selenka (Holothuroidea:Echinodermata). J. Exp. Zool. 281, 288–304. https://doi.org/10.1002/(SICI)1097-010X(19980701)281:4<288::AID-JEZ5>3.0.CO;2-K.

García-arrarás, J.E., Greenberg, M.J., 2001. Visceral regeneration in holothurians. Microsc. Res. Tech. 55, 438–451. https://doi.org/10.1002/jemt.1189.

García-Arrarás, J.E., Lázaro-Peña, M.I., Díaz-Balzac, C.A., 2018. Holothurians as a model system to study regeneration. In: Results and Problems in Cell Differentiation. Springer Verlag, pp. 255–283. https://doi.org/10.1007/978-3-319-92486-1_13.

González-Wangüemert, M., Valente, S., Aydin, M., 2015. Effects of fishery protection on biometry and genetic structure of two target sea cucumber species from the Mediterranean Sea. Hydrobiologia 743, 65–74. https://doi.org/10.1007/s10750-014-2006-2.

Günay, D., Emiroğlu, D., Tolon, T., Özden, O., Saygi, H., 2015. Farklı sıcaklıklarda deniz hıyarı (*Holothuria tubulosa*, Gmelin, 1788) genç bireylerinin büyüme ve yaşama oranı. Turk. J. Fish. Aquat. Sci. 15, 533–541. https://doi.org/10.4194/1303-2712-v15_2_41.

Hammond, L.S., 1982. Patterns of feeding and activity in deposit-feeding holothurians and echinoids (Echinodermata) from a shallow back-reef lagoon, Discovery Bay, Jamaica. Bull. Mar. Sci. 32, 549–571.

Hanafy, M.H., 2011. A study on the effect of the sea cucumber *Actinopyga mauritiana* (Echinodermata: Holothuroidea) on the sediment characteristics at El-Gemsha Bay, Red Sea coast, Egypt Mahmoud. Int. J. Environ. Sci. Eng. 2, 35–44.

Hartel, P.G., 2004. Microbial processes - environmental factors. In: Encyclopedia of Soils in the Environment. Elsevier Inc., pp. 448–455. https://doi.org/10.1016/B0-12-348530-4/00155-7

Hauksson, E., 1979. Feeding biology of stichopus tremulus, a deposit-feeding holothurian. Sarsia 64, 155–160. https://doi.org/10.1080/00364827.1979.10411376.

Hou, Y.R., Sun, Y.J., Gao, Q.F., Dong, S.L., Wen, B., Yu, H.B., 2017. Effect of the bioturbation derived from sea cucumber *Apostichopus japonicus* (Selenka) farming on the different occurrence forms of sedimentary inorganic carbon. Aquaculture 480, 108–115. https://doi.org/10.1016/j.aquaculture.2017.08.017.

Hulthe, G., Hulth, S., Hall, P.O.J., 1998. Effect of oxygen on degradation rate of refractory and labile organic matter in continental margin sediments. Geochem. Cosmochim. Acta 62, 1319–1328. https://doi.org/10.1016/S0016-7037(98)00044-1.

Inui, M., Itsubo, M., Iso, S., 1991. Creation of a new nonfeeding aquaculture system in enclosed coastal seas. Mar. Pollut. Bull. 23, 321–325. https://doi.org/10.1016/0025-326X(91)90694-N.

Işgören-Emiroğlu, D., Günay, D., 2007. The effect of sea cucumber Holothuria tubulosa G. 1788 on nutrient and organic matter contents of bottom sediment of oligotrophy and hypereutrophic shores. Fresenius Environ. Bull. 16, 290–294.

Kan-No, M., Kijima, A., 2003. Genetic differentiation among three color variants of Japanese sea cucumber *Stichopus japonicus*. Fish. Sci. 69, 806–812. https://doi.org/10.1046/j.1444-2906.2003.00690.x.

Kang, J.H., Kim, Y.K., Kim, M.J., Park, J.Y., An, C.M., Kim, B.S., Jun, J.C., Kim, S.K., 2011. Genetic differentiation among populations and color variants of sea cu-cumbers (*Stichopus japonicus*) from Korea and China. Int. J. Biol. Sci. 7, 323–332. https://doi.org/10.7150/ijbs.7.323.

Kang, K.H., Kwon, J.Y., Kim, Y.M., 2003. A beneficial coculture: charm abalone *Haliotis discus* hannai and sea cucumber *Stichopus japonicus*. Aquaculture 216, 87–93. https://doi.org/10.1016/S0044-8486(02)00203-X.

Kille, F.R., 1935. Regeneration in *Thyone briareus* Lesueur following induced autotomy. Biol. Bull. 69, 82–108. https://doi.org/10.2307/1537360.

Kim, M.J., Choi, T.J., An, H.S., 2008. Population genetic structure of sea cucumber, Stichopus japonicus in Korea using microsatellite markers. Aquacult. Res. 39, 1038–1045. https://doi.org/10.1111/j.1365-2109.2008.01962.x.

Kristensen, E., Penha-Lopes, G., Delefosse, M., Valdemarsen, T., Quintana, C.O., Banta, G.T., 2012. What is bioturbation? The need for a precise definition for fauna in aquatic sciences. Mar. Ecol. Prog. Ser. 446, 285–302.

Lee, S., Ferse, S., Ford, A., Wild, C., Mangubhai, S., 2017. Effect of sea cucumber density on the health of reef-flat sediments. Fiji's Sea Cucumber Fish. Adv. Sci. Improv. Manag. 54–61.

Leibson, N.L., 1992. Regeneration of digestive tube in holothurians *Stichopus japonicus* and *Eupentacta fraudatrix*. Monogr. Dev. Biol. 23, 51–61.

Li, Y., Wang, R., Xun, X., Wang, J., Bao, L., Thimmappa, R., Ding, J., Jiang, J., Zhang, L., Li, T., Lv, J., 2018. Sea cucumber genome provides insights into saponin biosynthesis and aestivation regulation. Cell Discov 4. https://doi.org/10.1038/s41421-018-0030-5.

Liu, Y., Li, F., Song, B., Sun, H., Zhang, X., Gu, B., 1996. Study on aestivating habit of sea cucumber *Apostichopus japonicus* Selenka: ecological characteristics of aestivation. J. Fish. Sci. China 3, 41–48.

Lohr, H.R., 2003. Genetic Variation Among Geographically Isolated Populations of the Commercially Important Sea Cucumber, Isostichopus Fuscus. The Eastern Pacific. Tesis fin grado.

Mackin, J.E., Swider, K.T., 1989. Organic matter decomposition pathways and oxygen consumption in coastal marine sediments. J. Mar. Res. 47, 681–716. https://doi.org/10.1357/002224089785076154.

MacTavish, T., Stenton-Dozey, J., Vopel, K., Savage, C., 2012. Deposit-feeding sea cucumbers enhance mineralization and nutrient cycling in organically-enriched coastal sediments. PloS One 7.

Mashanov, V.S., Zueva, O.R., Rojas-Catagena, C., Garcia-Arraras, J.E., 2010. Visceral regeneration in a sea cucumber involves extensive expression of survivin and mortalin homologs in the mesothelium. BMC Dev. Biol. 10 https://doi.org/10.1186/1471-213X-10-117.

Mercier, A., Battaglene, S.C., Hamel, J.F., 1999. Daily burrowing cycle and feeding activity of juvenile sea cucumbers *Holothuria scabra* in response to environmental factors. J. Exp. Mar. Biol. Ecol. 239, 125–156. https://doi.org/10.1016/S0022-0981(99)00034-9.

Michio, K., Kengo, K., Yasunori, K., Hitoshi, M., Takayuki, Y., Hideaki, Y., Hiroshi, S., 2003. Effects of deposit feeder *Stichopus japonicus* on algal bloom and organic matter contents of bottom sediments of the enclosed sea. Mar. Pollut. Bull. 47, 118–125. https://doi.org/10.1016/S0025-326X(02)00411-3.

Mirams, A.G.K., Treml, E.A., Shields, J.L., Liggins, L., Riginos, C., 2011. Vicariance and dispersal across an intermittent barrier: population genetic structure of marine animals across the Torres Strait land bridge. Coral Reefs 30, 937–949. https://doi.org/10.1007/s00338-011-0767-x.

Miyazaki, S., Ichiba, T., Reimer, J.D., Tanaka, J., 2014. Chemoattraction of the pearlfish *Encheliophis vermicularis* to the sea cucumber *Holothuria leucospilota*. Chemoecology 24, 121–126. https://doi.org/10.1007/s00049-014-0152-7.

Mohsen, M., Zhang, L., Sun, L., Lin, C., Liu, S., Wang, Q., Yang, H., 2020. A deposit-feeder sea cucumber also ingests suspended particles through the mouth. J. Exp. Biol 223 (24).

Moriarty, D.J.W., 1982. Feeding of Holothuria atra and *Stichopus chloronotus* on bacteria, organic carbon and organic nitrogen in sediments of the Great Barrier Reef. Mar. Freshw. Res. 33, 255—263. https://doi.org/10.1071/MF9820255.

Mosher, C., 1956. Observations on evisceration and visceral regeneration in the sea-cucumber, *Actinopyga agassizi* Selenka. Zool. N.Y. 41, 17—26.

Motokawa, T., 1984. Connective tissue catch in echinoderms. Biol. Rev. 59, 255—270. https://doi.org/10.1111/j.1469-185x.1984.tb00409.x.

Nelson, E.J., MacDonald, B.A., Robinson, S.M.C., 2012. The absorption efficiency of the suspension-feeding sea cucumber, *Cucumaria frondosa*, and its potential as an extractive integrated multi-trophic aquaculture (IMTA) species. Aquaculture 370—371, 19—25. https://doi.org/10.1016/j.aquaculture.2012.09.029.

Nevo, E., 2001. Evolution of genome-phenome diversity under environmental stress. Proc. Natl. Acad. Sci. U.S.A. 98, 6233—6240. https://doi.org/10.1073/pnas.101109298.

Ochoa-Chávez, J.M., Del Río-Portilla, M.Á., Calderón-Aguilera, L.E., Rocha-Olivares, A., 2018. Genetic connectivity of the endangered brown sea cucumber *Isostichopus fuscus* in the northern Gulf of California revealed by novel microsatellite markers. Rev. Mex. Biodivers. 89, 563—567. https://doi.org/10.22201/ib.20078706e.2018.2.2294.

Okada, A., Kondo, M., 2019. Regeneration of the digestive tract of an anterior-eviscerating sea cucumber, *Eupentacta quinquesemita*, and the involvement of mesenchymal-epithelial transition in digestive tube formation. Zool. Lett. 5, 21. https://doi.org/10.1186/s40851-019-0133-3.

Pawson, D.L., Caycedo, I.E., 1980. Holothuria (Thymiosycia) thomasi new species, a large Caribbean coral reef inhabiting sea cucumber (Echinodermata: Holothuroidea). Bull. Mar. Sci. 30, 454—459.

Pinn, W.S., Yasin, Z., Hwai, T.S., 2014. First record of Lissocarcinus orbicularis associated with sea cucumber, *Thelenota ananas*, in the north east of Celebes Sea. Malay. Nat. J. 66 (4), 368—375.

Pischedda, L., Poggiale, J.C., Cuny, P., Gilbert, F., 2008. Imaging oxygen distribution in marine sediments. The importance of bioturbation and sediment heterogeneity. Acta Biotheor. 56, 123—135. https://doi.org/10.1007/s10441-008-9033-1.

Purcell, S., 2004. Criteria for release strategies and evaluating the restocking of sea cucumbers. Adv. Sea Cucumber Aquac. Manag. 181—191.

Purcell, S.W., Conand, C., Uthicke, S., Byrne, M., 2016. Ecological roles of exploited sea cucumbers. In: Oceanography and Marine Biology: An Annual Review, pp. 367—386. https://doi.org/10.1201/9781315368597-8.

Ren, Y., Dong, S., Qin, C., Wang, F., Tian, X., Gao, Q., 2012. Ecological effects of co-culturing sea cucumber *Apostichopus japonicus* (Selenka) with scallop *Chlamys farreri* in earthen ponds. Chin. J. Oceanol. Limnol. 30, 71—79. https://doi.org/10.1007/s00343-012-1038-6.

Robinson, G., 2013. A bright future for sandfish aquaculture. World Aquacult. Soc. Mag. 18—24.

Schmutter, K., Nash, M., Dovey, L., 2017. Ocean acidification: assessing the vulnerability of socioeconomic systems in small Island developing States. Reg. Environ. Change 17, 973—987. https://doi.org/10.1007/s10113-016-0949-8.

Schneider, K., Silverman, J., Woolsey, E., Eriksson, H., Byrne, M., Caldeira, K., 2015. Potential influence of sea cucumbers on coral reef caco3 budget: a case study at one tree reef. J. Geophys. Res. 116 (G4).

Shiell, G.R., Knott, B., 2010. Aggregations and temporal changes in the activity and bioturbation contribution of the sea cucumber *Holothuria whitmaei* (Echinodermata: Holothuroidea). Mar. Ecol. Prog. Ser. 415, 127—139. https://doi.org/10.3354/meps08685.

Shumway, S.E., 1990. A review of the effects of algal blooms on shellfish and aquaculture. J. World Aquacult. Soc. 21, 65−104. https://doi.org/10.1111/j.1749-7345.1990.tb00529.x.

Skillings, D.J., Bird, C.E., Toonen, R.J., 2011. Gateways to Hawai 'i: genetic population structure of the tropical sea cucumber *Holothuria atra*. J. Mar. Biol. 2011, 1−16. https://doi.org/10.1155/2011/783030.

Slater, M.J., Carton, A.G., 2009. Effect of sea cucumber (*Australostichopus mollis*) grazing on coastal sediments impacted by mussel farm deposition. Mar. Pollut. Bull. 58, 1123−1129. https://doi.org/10.1016/j.marpolbul.2009.04.008.

Slater, M.J., Jeffs, A.G., Sewell, M.A., 2011. Organically selective movement and deposit-feeding in juvenile sea cucumber, *Australostichopus mollis* determined in situ and in the laboratory. J. Exp. Mar. Biol. Ecol. 409, 315−323. https://doi.org/10.1016/j.jembe.2011.09.010.

Sloan, N., von Bodungen, B., 1980. Distribution and feeding of the sea cucumber *Isostichopus badionotus* in relation to shelter and sediment criteria of the Bermuda platform. Mar. Ecol. Prog. Ser. 2, 257−264. https://doi.org/10.3354/meps002257.

Smith, G.N., 1971. Regeneration in the sea cucumber *Leptosynapta*. I. The process of regeneration. J. Exp. Zool. 177, 319−329. https://doi.org/10.1002/jez.1401770306.

So, J.J., Uthicke, S., Hamel, J.F., Mercier, A., 2011. Genetic population structure in a commercial marine invertebrate with long-lived lecithotrophic larvae: *Cucumaria frondosa* (Echinodermata: Holothuroidea). Mar. Biol. 158, 859−870. https://doi.org/10.1007/s00227-010-1613-3.

Soliman, T., Kanno, M., Kijima, A., Yamazaki, Y., 2012. Population genetic structure and gene flow in the Japanese sea cucumber *Apostichopus japonicus* across Toyama Bay, Japan. Fish. Sci. 78, 775−783. https://doi.org/10.1007/s12562-012-0509-1.

Swan, E.F., 1961. Seasonal evisceration in the sea cucumber, *Parastichopus californicus* (stimpson). Science 133, 1078−1079. https://doi.org/10.1126/science.133.3458.1078.

Tuwo, A., Yasir, I., Tresnati, J., Aprianto, R., Yanti, A., Bestari, A.D., Syafiuddin, Nakajima, M., 2019. Evisceration rate of sandfish *Holothuria scabra* during transportation. In: IOP Conference Series: Earth and Environmental Science, p. 12039. https://doi.org/10.1088/1755-1315/370/1/012039.

Uthicke, S., 2001. Nutrient regeneration by abundant coral reef holothurians. J. Exp. Mar. Biol. Ecol. 265, 153−170. https://doi.org/10.1016/S0022-0981(01)00329-X.

Uthicke, S., 1999. Sediment bioturbation and impact of feeding activity of *Holothuria* (Halodeima) *atra* and *Stichopus chloronotus*, two sediment feeding holothurians, at Lizard Island, great barrier reef. Bull. Mar. Sci. 64 (1), 129−141.

Uthicke, S., Benzie, J.A.H., 2000. Allozyme electrophoresis indicates high gene flow between populations of *Holothuria* (Microthele) *nobilis* (Holothuroidea: Aspidochirotida) on the Great Barrier Reef. Mar. Biol. 137, 819−825. https://doi.org/10.1007/s002270000393.

Uthicke, S., Karez, R., 1999. Sediment patch selectivity in tropical sea cucumbers (Holothurioidea: Aspidochirotida) analysed with multiple choice experiments. J. Exp. Mar. Biol. Ecol. 236, 69−87. https://doi.org/10.1016/S0022-0981(98)00190-7.

Uthicke, S., Purcell, S., 2004. Preservation of genetic diversity in restocking of the sea cucumber *Holothuria scabra* investigated by allozyme electrophoresis. Can. J. Fish. Aquat. Sci. 61, 519−528. https://doi.org/10.1139/f04-013.

Uthicke, S., 1994. Distribution patterns of two reef flat holothurians, Holothuria atra and Stichopus chloronotus, pp. 569−576.

Vergara-Chen, C., González-Wangüemert, M., Marcos, C., Pérez-Ruzafa, Á., 2010. Genetic diversity and connectivity remain high in *Holothuria polii* (Delle Chiaje 1823) across a coastal lagoon-open sea environmental gradient. Genetica 138, 895−906. https://doi.org/10.1007/s10709-010-9472-x.

Wolfe, K., Vidal-Ramirez, F., Dove, S., Deaker, D., Byrne, M., 2018. Altered sediment biota and lagoon habitat carbonate dynamics due to sea cucumber bioturbation in a high-pCO_2 environment. Global Change Biol. 24, 465−480. https://doi.org/10.1111/gcb.13826.

Wolkenhauer, S.M., Uthicke, S., Burridge, C., Skewes, T., Pitcher, R., 2010. The ecological role of holothuria scabra (Echinodermata: Holothuroidea) within subtropical seagrass beds. J. Mar. Biol. Assoc. U. K. 90, 215−223. https://doi.org/10.1017/S0025315409990518.

Wright, A.F., 2005. Genetic variation: polymorphisms and mutations. In: Encyclopedia of Life Sciences. John Wiley & Sons, Ltd., Chichester https://doi.org/10.1038/npg.els.0005005.

Yao, B., Hu, X., Bao, Z., Lu, W., Hu, J., 2007. Genetic variation in two sea cucumber (*Apostichopus japonicus*) stocks revealed by ISSR markers. Chin. J. Oceanol. Limnol. 25, 91−96. https://doi.org/10.1007/s00343-007-0091-z.

Yuan, X., Yang, H., Wang, L., Zhou, Y., Zhang, T., Liu, Y., 2007. Effects of aestivation on the energy budget of sea cucumber *Apostichopus japonicus* (Selenka) (Echinodermata: Holothuroidea). Acta Ecol. Sin. 27, 3155−3161. https://doi.org/10.1016/S1872-2032(07)60070-5.

Zamora, L.N., Jeffs, A.G., 2011. Feeding, selection, digestion and absorption of the organic matter from mussel waste by juveniles of the deposit-feeding sea cucumber, *Australostichopus mollis*. Aquaculture 317, 223−228. https://doi.org/10.1016/j.aquaculture.2011.04.011.

Zamora, L.N., Yuan, X., Carton, A.G., Slater, M.J., Mari ne, L., 2016. Role of deposit-feeding sea cucumbers in integrated multitrophic aquaculture: progress, problems, potential and future challenges. Rev. Aquacult. 10, 57−74. https://doi.org/10.1111/raq.12147.

Zhang, H., Xu, Q., Zhao, Y., Yang, H., 2016. Sea cucumber (*Apostichopus japonicus*) eukaryotic food source composition determined by 18s rDNA barcoding. Mar. Biol. 163 https://doi.org/10.1007/s00227-016-2931-x.

Zhang, X., Sun, L., Yuan, J., Sun, Y., Gao, Y., Zhang, L., Li, S., Dai, H., Hamel, J.F., Liu, C., Yu, Y., 2017. The sea cucumber genome provides insights into morphological evolution and visceral regeneration. PLoS Biol. 15, e2003790. https://doi.org/10.1371/journal.pbio.2003790.

Zhao, P., Yang, H.S., 2010. Selectivity of particle size by sea cucumber *Apostichopus japonicus* in different culture systems. Mar. Sci. 34, 11−16.

Zhao, Y., Chen, M., Su, L., Wang, T., Liu, S., Yang, H., 2013. Molecular cloning and expression-profile analysis of sea cucumber DNA (cytosine-5)-methyltransferase 1 and methyl-CpG binding domain type 2/3 genes during aestivation. Comp. Biochem. Physiol. B Biochem. Mol. Biol. 165, 26−35. https://doi.org/10.1016/j.cbpb.2013.02.009.

Zhou, Y., Yang, H., Liu, S., Yuan, X., Mao, Y., Liu, Y., Xu, X., Zhang, F., 2006. Feeding and growth on bivalve biodeposits by the deposit feeder *Stichopus japonicus* Selenka (Echinodermata: Holothuroidea) co-cultured in lantern nets. Aquaculture 256, 510−520. https://doi.org/10.1016/j.aquaculture.2006.02.005.

Ziemann, D.A., Walsh, W.A., Saphore, E.G., Fulton-Bennett, K., 1992. A Survey of water quality characteristics of effluent from Hawaiian aquaculture facilities. J. World Aquacult. Soc. 23, 180−191. https://doi.org/10.1111/j.1749-7345.1992.tb00767.x.

Sea cucumbers research in the Mediterranean and the Red Seas

4

4.1 Region under study

The Mediterranean Sea is nearly landlocked and usually denoted as a separated body of water. This interconnected body of saltwater is bounded on the south by North Africa, on the east by the Levant and the north by Southern Europe and Anatolia (Fig. 4.1). The area covered by the Mediterranean Sea is approximately 2,500,000 km^2, representing 0.7% of the global ocean surface, with an average depth of 1500 m. The deepest recorded point is 5267 m, which can be found in the Ionian Sea. The countries surrounding the Mediterranean Sea are Croatia, Bosnia and Herzegovina, Turkey, Spain, Lebanon, Monaco, Italy, Morocco, Slovenia, Montenegro,

FIGURE 4.1

The Mediterranean Sea and the Red Sea.

Credit: By O H 237 - Own work, CC BY-SA 4.0 via Wikimedia Commons.

Sea Cucumbers. https://doi.org/10.1016/B978-0-12-824377-0.00001-3

Albania, Greece, Syria, France, Palestine, Egypt, Libya, Tunisia, Algeria and Malta and Cyprus (island countries) (Coll et al., 2010; Pinardi et al., 2006).

The Red Sea lies between Africa and Arabia. The Gulf of Aden and Bab-el-Mandeb Strait lie in the south and are connected to the ocean (Figs. 4.1 and 4.2). The Gulf of Aqaba, the Sinai Peninsula and the Gulf of Suez are to the north, which leads to the Suez Canal. The Red Sea has a surface area that occupies an elongated depression. It ranges in width from 200 km in the north to 360 km in the south, with an average depth of 490 m. Its deepest point is 3040 m, and its area approximately 450,000 km^2 (Coleman, 1974; Drake and Girdler, 1964). The Red Sea is characterised as one of the saltiest seas in the world. This is likely because of high rates of evaporation, low precipitation, no rivers drain into the Red Sea and the connection to the Indian Ocean is limited. In the Bab el-Mandab Strait, salinity is 35 ppt, which is comparable with the global ocean average. However, the salinity increases steadily on a south-north axis to reach a maximum of 40.5 ppt in the north (Bailey, 2010). There are six countries surrounding the Red Sea, namely, Saudi Arabia, Yemen in the Eastern shore and Egypt, Sudan, Eritrea and Djibouti in the Western shore.

FIGURE 4.2

Map of the Red Sea.

Credit: By NormanEinstein CC BY-SA 3.0 via Wikimedia commons.

4.2 Biology and ecology of sea cucumber species

4.2.1 Sea cucumbers in the Mediterranean Sea

4.2.1.1 Holothuria tubulosa *(Gmelin, 1791)*

The sea cucumber *Holothuria tubulosa* is one of the most widespread species in the Mediterranean Sea. This species is found in depths between 5 and 100 m, inhabiting soft sediments, rocky substrata and seagrass beds. The body is brown and roughly cylinder, with a dorsal surface covered by brown and thick papillae (Fig. 4.3). This species does not have any Cuvierian tubules. The ventral surface is lighter in colour than the dorsal surface, and three longitudinal rows of tube feet are present. Ossicles were found in the form of buttons, rods and plates (Aydin and Erkan, 2015; Dereli et al., 2016; Ocaña and Sánchez Tocino, 2005).

The reproductive cycle of the sea cucumber *H. tubulosa* at different regions of the Mediterranean Sea was reported. Despalatović et al. (2004) analysed the reproductive cycle of the sea cucumber *H. tubulosa* in Kaštela Bay (Adriatic Sea) from July 1994 to August 1995. Spawning occurs from July to September during the warm season, with a surface water temperature of 22–26°C. Individuals were in the resting phase from October to January and without gonads (Despalatović et al., 2004). Likewise, *H. tubulosa* collected from the Southwestern Mediterranean, specifically from the Algiers coastal area, showed an annual reproductive cycle taking place from June to August. The maturation occurs in spring and the spawning at the end of summer until the middle of autumn followed by a winter resting phase (Mezali and Soualili, 2014). Also, in the Dardanelles Strait, *H. tubulosa* reproduction was observed between August and September, and their gonadosomatic index reached maximum values in July (Dereli et al., 2016). In the Aegean Sea, the reproductive biology of *H. tubulosa* was observed in June, July, August and September (Aydin and Erkan, 2015). *H. tubulosa* collected off Ischia Island (Italy) was observed to spawn at the end of summer (August to September), and gonadal growth starts by

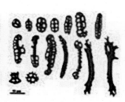

FIGURE 4.3

The sea cucumber *Holothuria tubulosa.*

Credit: Aydin, M., 2016. Sea cucumber (Holothuroidea) species of Turkey Türkiye Denizlerindeki deniz hıyarı (Holothuroidea) Türleri, Turk. J. Maritime Mar. Sci. 2 (1), 49–58.

mid-winter (Bulteel et al., 1992). Collectively, the sea cucumber *H. tubulosa* was observed to spawn during the warm season in the Mediterranean Sea.

In the habitat, the natural spawning of the sea cucumber *H. tubulosa* was observed and illustrated. In the Alboran Sea, in August 2003, during the full moon on 11th and 12th of August, *H. tubulosa* sea cucumbers were observed in spawning position at different times in the afternoon, from 4 to 6 h solar time, with a range of water temperature from 24 to 25°C. Individuals were observed in a spawning position. The body of the sea cucumber was in copra position (i.e. about the last half of the body was in contact with the substratum and the rest of the body stays upright) (Ocaña and Sánchez Tocino, 2005). Also, natural spawning of the sea cucumber *H. tubulosa* was observed at 1.5—2 m, close to shore between 27 June and 3 July 2003 at 15:00 in Costa Brava, Mediterranean Sea (Valls, 2004).

Observations revealed that environmental factors could affect the growth of *H. tubulosa* and its gonadal development (Bulteel et al., 1992). *H. tubulosa* could grow more efficiently in areas of low water turbulence, where seagrass and detrital food are accessible (Bulteel et al., 1992). Also, Vafeiadou et al. (2010) concluded that *H. tubulosa* population might thrive only in areas with high organic inputs. Furthermore, the feeding rate of *H. tubulosa* was correlated with the season (i.e. an increase of water temperature leads to an increase of feeding rate) and with the nycthemeral cycle (i.e. individuals being more active at night). The feeding rate ranges over one order of magnitude (from 0.049 to 0.489 g dw sediment h^{-1}) in a Mediterranean seagrass bed off Ischia Island, Italy, depending on the size of individuals and environmental conditions (Coulon and Jangoux, 1993).

4.2.1.2 Holothuria sanctori *(Delle Chiaje, 1823)*

The sea cucumber *Holothuria sanctori* was found in the Mediterranean coast of Egypt on the rocky substratum and rocky cavities at 5—22 m depth (Moussa and Wirawati, 2018). Also, *H. sanctori* was found in the Aegean Sea within 0—20 m depth (Aydin and Erkan, 2015). Body colouration is brown and dark brown (Fig. 4.4). The outer surface is leathery and thick. The dorsal surface has papillae,

FIGURE 4.4

The sea cucumber *Holothuria sanctori*.

Credit: Aydin, M., 2016. Sea cucumber (Holothuroidea) species of Turkey Türkiye Denizlerindeki deniz hıyarı (Holothuroidea) Türleri, Turk. J. Maritime Mar. Sci. 2 (1), 49—58.

and the ventral surface has tube feet arranged in three longitudinal lines. The ossicles were found in the form of buttons, tables, plates and rods (Moussa and Wirawati, 2018).

The reproductive period of *H. sanctori* was observed in June, July, August and September in the Aegean Sea (Aydin and Erkan, 2015). Also, the reproductive biology of the sea cucumber *H. sanctori* from the Algerian coastline in the southwestern Mediterranean was reported, where *H. sanctori* had a winter sexual resting phase followed by maturation during spring and spawning in summer (Mezali et al., 2014).

4.2.1.3 Holothuria arguinensis *(Koehler & Vaney, 1906)*

The sea cucumber *Holothuria arguinensis* is a holothurian species that does not take refuge in crevices and remains visible on sandy or seagrass beds. It inhabits the intertidal zone and can stand periods of drought during the low tides (González-Wangüemert and Borrero-Pérez, 2012). The presence of *H. arguinensis* was reported from the Algerian coastal waters. Individuals of this species were taken from the Tamentefoust area at 4.5 m depth and examined. The length of the live animals was about 350 mm, and their wet body weight was 270 g (Mezali and Thandar, 2014). The body is rigid, roughly cylindrical and dark brown. The dorsal side includes regularly arranged conical warts, in two double rows, constituting an almost continuous border with 16 on the left and 15 on the right, each terminating in a white papilla (Fig. 4.5). Other papillae are also white but borne on much smaller warts. Also, the ventral surface has a light brown colour with scattered tube feet. The ventral mouth has 20 tentacles coloured lemon-yellow. Ossicles were found in the form of buttons, rods and plates. The ossicles in the dorsal surface differ in shape from the ventral surface (Mezali and Thandar, 2014).

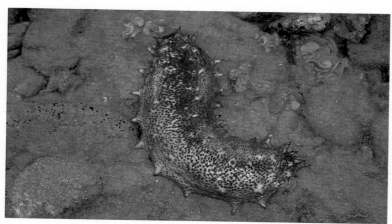

FIGURE 4.5

The sea cucumber *Holothuria arguinensis*.

Credit: Daniel S. CC BY-SA 3.0 via Wikimedia commons.

The reproductive biology of the sea cucumber *H. arguinensis* was observed in the southern Iberian Peninsula. Spawning occurs during summer-autumn, and a recovery phase was found in winter. This pattern was correlated with temperature and photoperiod in *H. arguinensis* (Marquet et al., 2017). Gametogenesis period, better growth and gonadal production are likely influenced by the particular features of each location, such as food availability and tidal stress (Marquet et al., 2017).

4.2.1.4 Stichopus regalis *(Cuvier, 1817)*

The sea cucumber *Stichopus regalis* is a common species in the Mediterranean Sea. This species can be found at depths of 5–800 m; 18–20 tentacles surround its mouth, and the body colouration is orange-yellow (Fig. 4.6) (Aydin, 2016). It is the most expensive seafood product in the Catalan market (Ramón et al., 2010). Ocaña et al. (1982) found *S. regalis* between 80 and 290 m in the Alboran Sea when sampling was carried out between 0 and 400 m depth (Ocaña et al., 1982). Aydin and Erkan (2015) found the sea cucumber *S. regalis* at depths of 0–20 m, and the reproductive period was observed in June, July, August and September (Aydin and Erkan, 2015). Furthermore, Ramón et al. (2010) examined the population of the sea cucumber *S. regalis* in the Menorca and Mallorca continental shelf from 2001 to 2009. Individual's abundance was highest between 100 and 299 m depth, and sizes were largest between 50 and 299 m depth (Ramón et al., 2010). Furthermore, Massutí and Renones (2005) found that the sea cucumber *S. regalis* was among the most commercial species captured in the survey in spring and autumn between 100 and 200 m in the fishing grounds off the Balearic Islands (Massutí and Renones, 2005). In 2015, a study carried out on *S. regalis* in Mediterranean populations from the Spanish coast found adverse effects of overexploitation in one population from Catalonia (Maggi and González-Wangüemert, 2015).

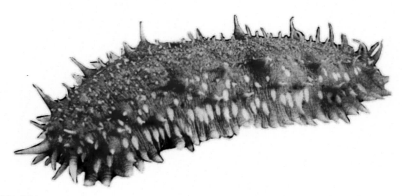

FIGURE 4.6

The sea cucumber *Stichopus regalis*.

Credit: Aydin, M., 2016. Sea cucumber (Holothuroidea) species of Turkey Türkiye Denizlerindeki deniz hıyarı (Holothuroidea) Türleri, Turk. J. Maritime Mar. Sci. 2 (1), 49–58.

4.2.1.5 Holothuria poli *(Delle Chiaje, 1824)*

The sea cucumber *Holothuria poli* is a common sea cucumber species in the Mediterranean and has recently become a target species for exploitation. The body is cylindrical with a dark brown, and the ventral surface has a light colour. The dorsal side has yellowish to white scattered papillae, whereas the ventral surface has white tube feet (Fig. 4.7). The mouth has 20–22 peltate tentacles. The ossicles were found in the form of buttons, tables, plates and rods (Eissa et al., 2017; Moussa and Wirawati, 2018).

The sea cucumber *H. poli* was found on a rocky and sandy substratum of 2–12 m depth in the Mediterranean Sea coast of Egypt (Moussa and Wirawati, 2018). Also, during 2008–12, the sea cucumber *H. poli* was collected from the Aegean Sea. The sea cucumber was distributed within 0–20 m depth, and their reproductive period was observed in June, July, August and September (Aydin and Erkan, 2015).

The spawning period of *H. poli* was observed from June to August at Gera Gulf of Lesvos Island (Bardanis and Batjakas, 2018). Additionally, the reproduction biology of the sea cucumber *H. poli* was studied in the Aegean Sea during July to October 2018. The spawning stage was observed in July, where the gonads were mature, and the post-spawning stage was observed in October. The spawning was correlated with the water temperature. Therefore, the spawning period of *H. poli* at the eastern coasts of the Aegean Sea starts in July and ends in October (Tolon and Engin, 2019).

Furthermore, the sea cucumber *H. poli* was collected from Kristel Bay at Ain Franine in Oran Province, Algeria, for reproductive biology examination (Slimane-Tamacha et al., 2019). The size at first sexual maturity within the entire population was 135 mm. Five macroscopic and microscopic sexual maturity stages have been identified in the gonadal tubules: recovery (I), growing (II), early mature (III), mature (IV) and spent (V). The maturation of the gonads (stages III and IV) occurs from March until May. From May to July, the entire sampled population is at full sexual maturity, and it is only in July that spawning begins, which extends up to September. The period of non-reproductive activity was found between October and November (Slimane-Tamacha et al., 2019).

FIGURE 4.7

The sea cucumber *Holothuria poli*.

Credit: Aydin, M., 2016. Sea cucumber (Holothuroidea) species of Turkey Türkiye Denizlerindeki deniz hıyarı (Holothuroidea) Türleri, Turk. J. Maritime Mar. Sci. 2 (1), 49–58.

Furthermore, among seven locations in the Mediterranean Sea, the sea cucumber *H. poli* from Turkey showed the highest genetic diversity, locating the origin of *H. poli* in Turkish waters and suggesting the capacity to recover from population decline through supplementation from other areas (Valente et al., 2015). Furthermore, *H. poli* sampled from the eastern and western Mediterranean coastal waters of Tunisia showed low genetic variability, which suggests a population structure consistent with separation by Mediterranean Sea basins that might reflect different local biogeographical zones (Gharbi and Said, 2011). Fishery pressure could be reducing the size of the individuals and decreasing the genetic diversity of sea cucumber stocks (González-Wangüemert et al., 2015). Better management of the sea cucumber fisheries will help in conserving the genetic diversity of the species (González-Wangüemert et al., 2015).

4.2.1.6 Holothuria mammata *(Grube, 1840)*

The sea cucumber *Holothuria mammata* is one of the new target species for exploitation in the Mediterranean. It prefers the rocky bottom, and its body colouration is more purple and darker than *H. tubulosa* (Fig. 4.8). Also, *H. mammata* has larger papillae than *H. tubulosa*. Cuvierian tubules exist, but there are few. The ossicles were found in the form of tables, buttons, rods and plates, and their size is bigger than those found in the sea cucumber *H. tubulosa* (Aydin and Erkan, 2015). In the Aegean Sea, *H. mammata* were found within 0−20 m depth (Aydin and Erkan, 2015).

The reproductive biology of *H. mammata* was investigated in the southern Iberian Peninsula by Marquet et al. (2017). *H. mammata* living in a narrow latitudinal range have the same general reproductive pattern, with a spawning period during summer-autumn and a recovery phase in winter. The authors found differences in size/weight, gonadal production and maturity stages between locations and attributed it to the differences in food availability and tidal stress of the sites (Marquet et al., 2017). Also, in the Aegean Sea, the reproductive period of the sea cucumber *H. mammata* was observed in June, July, August and September (Aydin and Erkan, 2015).

FIGURE 4.8

The sea cucumber *Holothuria mammata*.

Credit: Aydin, M., 2016. Sea cucumber (Holothuroidea) species of Turkey Türkiye Denizlerindeki deniz hıyarı (Holothuroidea) Türleri, Turk. J. Maritime Mar. Sci. 2 (1), 49−58.

Furthermore, the population of the sea cucumber *H. mammata* has the lowest genetic diversity compared with populations in the Macaronesian islands, Western Mediterranean. This observation was ascribed to the restriction of gene flow, despite the long duration of its planktonic larval stage (Borrero-Pérez et al., 2011). Moreover, nine novel microsatellites were recommended for future genetic monitoring on *H. mammata*, tracking the fishery effects in the *H. mammata* populations, and possibly safeguarding their evolutionary potential (Henriques et al., 2016).

4.2.1.7 Other species

In the Mediterranean Sea coast of Egypt, two sea cucumber species were reported during a survey conducted between 2005 and 2007. Two thousand and 15 specimens were collected along the Alexandria coast. Among them, the sea cucumber *Holothuria arenicola* was the most abundant species with a percentage of 89% over the sea cucumber *Bohadschia argus*, with dominance to males (Razek et al., 2010). Individuals lacking gonads were found in January and December (Razek et al., 2010). Also, during 2008—12, the sea cucumber *Holothuria forskali* was collected and identified from the Aegean Sea. It was distributed within 0—20 m depth, and its reproductive period was observed in June, July, August and September (Aydin and Erkan, 2015) (Figs. 4.9—4.11; Table 4.1). Furthermore, Mezali et al. (2020) observed the sea cucumber *Paraleptopentacta tergestina* n. comb. at the depth of 60 m in the Mostaganem region, west of the Algerian coast (Mezali et al., 2020).

4.2.2 Sea cucumbers in the Red Sea

The Red Sea coast of Egypt extends over about 1500 km in length, including the Gulfs of Suez and Aqaba (Gvirtzman, 1977). Overexploitation of sea cucumber

FIGURE 4.9

The sea cucumber *Holothuria arenicola*.

Credit: By Philippe Bourjon via Wikimedia Commons. Licensed under CC BY-SA 3.0.

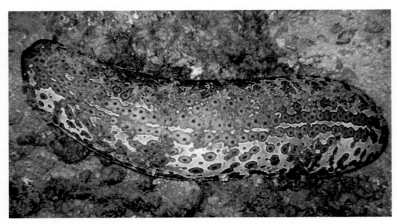

FIGURE 4.10

The sea cucumber *Bohadschia argus*.

Credit: By KKPCW via Wikimedia Commons, CC BY-SA 4.0.

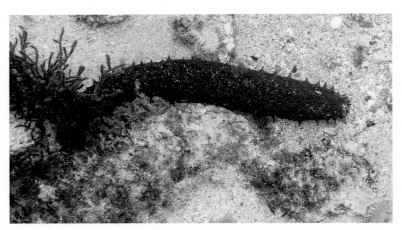

FIGURE 4.11

The sea cucumber *Holothuria froskali*

Credit: By Diego Delso, CC BY-SA 4.0.

species was reported from the Red Sea coast of Egypt. A survey was conducted along the Red Sea coast and offshore islands of Egypt from 2002 to 2003 in 116 sites (Lawrence et al., 2004). A total of 22 species were identified, of which seven species are commercially important (Table 4.2). The highest density found was for the sea cucumber *Actinopyga mauritiana*, which has a low commercial value (Ahmed and Lawrence, 2007; Lawrence et al., 2004). The sea cucumber *Bohadschia cousteaui*

Table 4.1 Habitat preference of sea cucumber species in the Mediterranean Sea.

Species	Habitat	Reference
Holothuria tubulosa	Frequently found in the Aegean coast up to 50 m depth, on the bedrock covered algae, meadows of *Posidonia* and other phanerogams, sandy, muddy gravelly and stony habitats. Reproduction occurs in summer.	Aydin and Erkan (2015), Dereli et al. (2016)
Holothuria poli	Frequently found between 0 and 100 m depth. Mostly prefers soft muddy substrates and *Zostera* meadows. The reproductive period was observed in June, July, August and September.	Aydin and Erkan (2015), Eissa et al. (2017)
Holothuria mammata	Found in shallow waters and up to 200 m depth. Mostly prefers stony and rocky substrates and occasionally sandy bottoms (Özgür Özbek, 2013). The reproductive period was observed in June, July, August and September.	Aydin and Erkan (2015)
Holothuria sanctori	It is common in the Aegean Sea. Mostly prefers darker places between rocks, caves and hollows. Found in shallow waters up to 50 m. The reproductive period was observed in Summer.	Aydin (2008), Aydin and Erkan (2015)
Parastichopus regalis	It prefers muddy, sandy-muddy and gravelly substrates and found between 5 and 800 m depth.	Aydin (2016), Aydin and Erkan (2015)
Holothuria (Theelothuria) hamata	It is common in the Mediterranean Sea, Turkey, Iskenderun Bay, at 30 m depth.	Aydin (2019)
Holothuria forskali	It is found at 0—20 m depth, and its reproductive period was observed in June, July, August and September.	Aydin and Erkan (2015)
Synaptula reciprocans	Mostly found on hard surface and sandy floor.	Aydin (2016)
Stereoderma kirschbergii	Found at a depth of 30 m in the Black Sea.	Sezgin et al. (2007)

was also among the most abundant species in the Red Sea (Ahmed, 2009). Another survey was conducted in 2007 at four sites of the Red Sea (Elgemsha Bay, Palm Beach, Nabq and Nuweiba). The authors reported that there was limited or no evidence of stock recovery, likely because of the slow regeneration of sea cucumber and the illegal fishing in some places in Egypt (Ahmed and Lawrence, 2007). El-Ganainy et al. (2006) reported that from 1999 to 2005, the mean size of the sea cucumber *Holothuria atra* collected decreased from 22.39 ± 6.24 cm to 15.33 ± 5.19 cm and that of *Holothuria leucospilota* decreased from 38.18 ± 11.30 to 35.73 ± 8.58. During the same period, the exploitation ratio of

Table 4.2 Habitat preference for sea cucumbers reported from the Red Sea.

Species	Habitat	References
Actinopyga crassa	It was found at a depth of 3–20 m in seagrass beds and, sometimes, around coral patches as well as the sandy lagoon.	Ahmed (2009), Hellal, (2010), Lawrence et al. (2004)
Actinopyga echinites	It was found in dead and live coral habitats, sandy patches and reef as well as seagrass beds.	Melek et al. (2012), Hasan (2019), Hellal (2010)
Actinopyga mauritiana	It was found in the seagrass beds, sand and rarely found in corals. Mostly found at a depth of 5–10 m and can be found at depth 10–20 m and 20–30 m.	Lawrence et al. (2004), Ahmed (2009), Hasan (2019), El-Naggar et al. (2008), Hellal (2010)
Actinopyga miliaris	It was found in shallow water on a muddy-sandy bottom.	Hasan (2019)
Bohadschia cousteaui	It was found in subtidal and shallow water on sandy and rocky substrates, mostly at a depth less than 20 m.	Lawrence et al. (2004), Ahmed (2009)
Bohadschia mitsioensis	It was found in sandy, rocky habitats seagrass and dead corals.	Hellal (2010)
Bohadschia steinitzi	It was found at depths of 5–25 m in the sandy lagoon but not in the coral area.	Hellal (2010)
Bohadschia tenuissima	It was found in seagrass beds and sandy lagoons at a depth of 5–15 m.	Ahmed (2009), Hellal (2010), Lawrence et al. (2004)
Bohadschia vitiensis	It was found in sandy habitats in seagrass and corals.	Hasan (2019), Hellal (2010), Lawrence et al. (2004)
Bohadschia marmorata	It was found burrowing in sandy-muddy sediment mostly at a depth less than 20 m along the Red Sea coast and Gulf of Aqaba.	Ahmed (2009)
Holothuria albiventer	It was found in the Red Sea coast of Egypt at a depth less than 10 m, inhabiting seagrass beds, coral reef, subtidal rocky and sandy beaches, hiding under shelters (rocks).	Ahmed (2009)
Holothuria atra	It was found in sandy and muddy areas in the seagrass beds, and corals at a depth less than 20 m and can be found at depths of 0.5–6 m.	Ahmed (2009), El-Ganainy et al. (2006), Hasan (2019), Hellal (2010), Lawrence et al. (2004)

Table 4.2 Habitat preference for sea cucumbers reported from the Red Sea.—*cont'd*

Species	Habitat	References
Holothuria arenicola	It was found in sandy beaches and rocky habitats and dead and live corals.	Hellal (2010)
Holothuria albofusca	It was found in sandy patches or under rocks at a depth of 0.5—3 m.	Hellal (2010)
Holothuria coluber	It was reported from the Red Sea coast and offshore islands of Egypt.	Lawrence et al. (2004)
Holothuria conusalba	It was found in sandy and muddy beaches in the subtidal region and the seagrass areas.	Hellal (2010)
Holothuria crosnieri	It was found between dead and live coral patches and sandy substrates at a depth of 20—25m.	Hellal (2010)
Holothuria edulis	It was found at depth more than 30 m distributed in rubbles and live and dead corals.	Ahmed (2009), Hellal (2010), Lawrence et al. (2004)
Holothuria fuscogilva	It was found in the seagrass beds, corals and subtidal rocky beaches hiding under rocks at a depth of 20—30 m and can be found at 5—10 m and 10—20 m depths.	Ahmed (2009), El-Naggar et al. (2008), Hasan (2019), Lawrence et al. (2004)
Holothuria hilla	It was found in sandy lagoons between rocks and corals.	Lawrence et al. (2004)
Holothuria impatiens	Mostly found at a depth less than 25 m in coral, reef flat and sandy beaches.	Ahmed (2009), Hellal (2010)
Holothuria leucospilota	It was found at a depth less than 20 m in seagrass beds, sandy-muddy sediment and hiding in corals.	Ahmed (2009), El-Ganainy et al. (2006)
Holothuria nobilis	Mostly found in reef flat and corals at 5—10 m depth.	Ahmed (2009), El-Naggar et al. (2008), Lawrence et al. (2004)
Holothuria pardalis	It was reported from the Red Sea coast and offshore islands of Egypt.	Lawrence et al. (2004)
Holothuria papillifera	It was found in sandy and seagrass beds.	Hellal (2010)
Holothuria sucosa	It was found living between live and dead corals and rocky patches.	Hellal (2010)

Continued

Table 4.2 Habitat preference for sea cucumbers reported from the Red Sea.—*cont'd*

Species	Habitat	References
Holothuria poli	It was found in the rocky and sandy substratum of 2–12 m depth.	Melek et al. (2012), Moussa and Wirawati (2018), Omran and Khedr (2015)
Holothuria rigida	It was reported from the Red Sea coast and offshore islands of Egypt.	Lawrence et al. (2004)
Holothuria sanctori	It was found in the rocky substratum and rocky cavities at 5–22 m depth.	Moussa and Wirawati (2018)
Holothuria thomasi	It was reported from Hurghada, Red Sea, Egypt.	El Barky et al. (2016)
Holothuria scabra	It was found in the seagrass beds and burrowing in sandy-muddy sediment at a depth of 5–10 m and can be found at 10–25 m.	Ahmed (2009); Hasan (2019), Hellal (2010), Lawrence et al. (2004)
Holothuria sp.	It was reported from the Red Sea coast and offshore islands of Egypt.	Lawrence et al. (2004)
Holothuria spinifera	It was found in sandy and seagrass patches also found in rocky substrates and live and dead corals.	Hellal (2010)
Stichopus variegatus	It was found in sandy habitats and seagrass at depths of 3–35 m.	Hellal (2010)
Labidodemas rugosum	It was found buried in the sand under dead corals and rocks at depth up to 30 m.	Hellal (2010)
Ohshimella ehrenbergii	It was found in the rocky and dead coral areas in both intertidal and subtidal zones as well as live corals at a depth of 3 m.	Hellal (2010)
Pearsonothuria graeffei	It was found in coral at a depth of 5–10 m and can be found at 10–20 m.	Ahmed (2009), Lawrence et al. (2004)
Polyplectana kefersteini	It was found in sandy lagoons and patches in the reef flat.	Hellal (2010)
Stichopus hermanni	It was reported from the Red Sea coast and Gulf of Aqaba at a depth less than 20 m in muddy-sand sediment, seagrass beds and rubbles.	Ahmed (2009), Hasan (2019)
Stichopus horrens	It was found in the seagrass beds. Also, it is found in sandy lagoons, live corals and coral patches. Mostly found at a depth of 5–10 m and can be found at 10–20 m.	Hasan (2019), Hellal (2010), Lawrence et al. (2004)

Table 4.2 Habitat preference for sea cucumbers reported from the Red Sea.—*cont'd*

Species	Habitat	References
Synapta maculate	It was found at a depth of 5—20 m in seagrass and sandy lagoons.	Ahmed (2009), Hellal (2010)
Synaptula reciprocans	It was found in subtidal areas in seagrass beds, sandy lagoons and sometimes in dead corals.	Ahmed (2009)
Synaptula sp.	It was reported from the Red Sea coast and Gulf Aqaba of Egypt.	Lawrence et al. (2004)

H. atra increased by about 31.14%, and by about 15% for *H. leucospilota* (El-Ganainy et al., 2006). Recently, Hasan (2019) reported that the population of sea cucumbers in the Abu Ghosoun area in the Red Sea coast of Egypt has decreased from 13 species in 2000 and 2006 to only 7 species in 2016 (Fig. 4.12) (Hasan, 2019). The population density has been severely affected as a result of the uncontrolled fishery and the absence of management. Over 16 years, 82.6% of the sea cucumber populations were lost (Hasan, 2019). The population of the sea cucumber *Thelenota ananas* completely disappeared, whereas the population of the sea cucumber *H. atra* is in high density (Fig. 4.12) (Hasan, 2019).

The Red Sea of Saudi Arabia extends over 2000 km. In 2006, 12 sea cucumber species were recorded from the coast of Saudi Arabia, including *Holothuria*

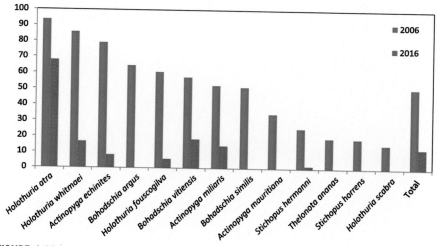

FIGURE 4.12

A comparison between the abundance of sea cucumber species in 2006 and 2016 at Abu Ghosoun area, Red Sea.

Credit: Hasan, M.H., 2019. Destruction of sea cucumber populations due to overfishing at Abu Ghosoun area, Red Sea. J. Basic Appl. Zool. 80. https://doi.org/10.1186/s41936-019-0074-6 licensed under CC-BY 4.0.

fuscogilva, *Holothuria nobilis*, *Holothuria scabra*, *H. atra*, *Holothuria edulis*, *H. leucospilota*, *Holothuria fuscopunctata*, *Actinopyga echinites*, *A. mauritiana*, *Stichopus hermanni Bohadschia vitiensis* and *Pearsonothuria graeffei* (Hasan, 2009).

From the Red Sea of Eretria, 16 sea cucumber species were identified, mostly found at depths of 6 and 15 m. The commercial species *H. atra*, *H. edulis*, *Actinopyga miliaris* and *A. mauritiana* were found mostly on corals and sandy substrates. *H. scabra* was found in muddy areas and seagrass beds. *H. atra* was the most abundant species (Kalaeb et al., 2008).

From the Bab el-Mandab area (the entrance of the Red Sea) of Yemen, a total of 28 species were collected from Bab el-Mandab, El-Khokha, El-Makha harbour, Gulf of Aden and mangrove area. Bab el-Mandab area had the highest species diversity (Hellal, 2010). Sandy habitats have the highest number of sea cucumber species followed by corals, seagrass beds and rocky habitats (Hellal, 2010).

4.2.2.1 Holothuria scabra *(Jaeger, 1833)*

The sea cucumber *H. scabra*, also known as sandfish, has an oval body. It has black dorsal surface and sometimes whitish to grey, with small papillae. The ventral surface has a whitish colour lighter than the dorsal surface (Fig. 4.13). This species is found covered by sediment. Its ventral mouth is surrounded by 20 tentacles grey short, stout and grey. Furthermore, the ossicles were found in the shape of rods in the tentacles. Tables, buttons and rods were found in the dorsal body wall (Ahmed, 2009).

The sea cucumber *H. scabra* was recorded only on sandy substrates in the Red Sea of Saudi Arabia (Hasan, 2009). From the Red Sea coast of Egypt, the sea cucumber *H. scabra* is reported from the shallow water at a depth of 5−10 m, and it

FIGURE 4.13

The sea cucumber *Holothuria scabra*.

Credit: Ahmed, M., 2009. Morphological, Ecological and Molecular Examination of the Sea Cucumber Species along the Red Sea Coast of Egypt and Gulf of Aqaba. The University of Hull.

is rarely found at depths of 10—25 m as it hides in seagrass, coral patches, rubble or buried in the sandy-muddy ground (Ahmed, 2009). Dar (2004) highlighted the importance of holothurians, including *H. scabra* in providing a favourable environment for benthos and fauna together with their positive effects on benthic organic matter and nutrient recycling in the Red Sea. Additionally, the natural population of the sea cucumber *H. scabra* at Abu Rhamada Island in the Red Sea was observed from 1991 to 2003. The population had a significant decrease in the density, abundance and biomass by 2003 because of the overfishing that began in June 2001, which can negatively affect the habitat in the Red Sea (Hasan, 2005).

4.2.2.2 Holothuria fuscogilva *(Cherbonnier, 1980)*

The sea cucumber *H. fuscogilva*, also known as the white teatfish, is usually found buried in the sand. The body is sub-oval, thick and rigid. The dorsal surface is brown and has whitish spots with large papillae, whereas the ventral surface has a lighter and a whitish colour (Fig. 4.14). The ventral mouth of this species is surrounded by 20 stout and grey tentacles. The ossicles were found in the tentacles in the form of rods. Tables and buttons ossicles were found in the ventral body wall, and tables were found in the dorsal body wall (Ahmed, 2009).

In the Red Sea of Saudi Arabia, the sea cucumber *H. fuscogilva* was found in rocky and dead coral habitats (Hasan, 2009). In Egypt, this species has been reported from the Red Sea coast of Egypt and Gulf of Aqaba (Lawrence et al., 2010) at 10 m deep between coral reefs, seagrass, sandy and subtidal rocky beaches (Ahmed, 2009).

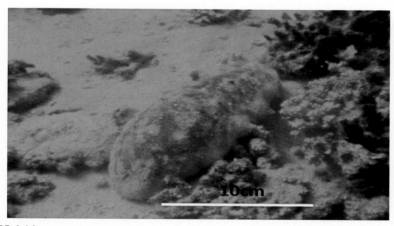

FIGURE 4.14

The sea cucumber *Holothuria fuscogilva*.

Credit: Ahmed, M., 2009. Morphological, Ecological and Molecular Examination of the Sea Cucumber Species along the Red Sea Coast of Egypt and Gulf of Aqaba. The University of Hull.

4.2.2.3 Holothuria nobilis *(Selenka, 1867)*

The sea cucumber *H. nobilis*, also known as black teatfish, is found covered by sand. The body is sub-oval, thick and rigid. The dorsal surface is black and has large papillae. The ventral surface is greyish and lighter in colour than the dorsal surface (Fig. 4.15). Twenty grey tentacles surround the ventral mouth. The Cuvierian tubules exist. The ossicles were found in the tentacles in the form of rods and tables. In the ventral and dorsal body wall, the ossicles were found in the form of tables, ellipsoids and plates (Ahmed, 2009).

In the Red Sea of Saudi Arabia, the sea cucumber *H. nobilis* was recorded in rocky and dead coral habitats (Hasan, 2009). This species is reported from the Red Sea coast of Egypt and the Gulf of Aqaba (Lawrence et al., 2010). In Egypt, it can be found across reef flats or seagrass at shallow water 15 m depth (Ahmed, 2009).

4.2.2.4 Stichopus hermanni *(Semper, 1868)*

The body of the sea cucumber *S. hermanni* is yellow to greenish with black spots. The dorsal surface has eight longitudinal rows of irregular conical warts, with smaller papillae. The ventral surface is lighter than the dorsal surface and has podia yellow to pink in colour (Fig. 4.16). The ventral mouth is surrounded by 20 tentacles yellowish in colour. The ossicles were found in the tentacles in the form of rods. In the ventral and dorsal body wall, the tentacles were found in the form of tables, rosettes and C-shaped rods (Ahmed, 2009).

The sea cucumber *S. hermanni* was reported from the Red Sea coast of Egypt and the Gulf of Aqaba (Lawrence et al., 2010). This species inhabits sandy mud grounds, seagrass beds and debris at depth reaching to 20 m (Ahmed, 2009).

FIGURE 4.15

The sea cucumber *Holothuria nobilis*.

Credit: Ahmed, M., 2009. Morphological, Ecological and Molecular Examination of the Sea Cucumber Species along the Red Sea Coast of Egypt and Gulf of Aqaba. The University of Hull.

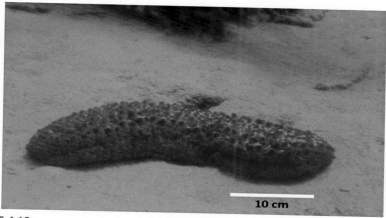

FIGURE 4.16

The sea cucumber *Stichopus hermanni*.

Credit: Ahmed, M., 2009. Morphological, Ecological and Molecular Examination of the Sea Cucumber Species along the Red Sea Coast of Egypt and Gulf of Aqaba. The University of Hull.

4.2.2.5 Actinopyga mauritiana *(Quay and Gaimard, 1833)*

The body of the sea cucumber *A. mauritiana* is cylindrical and elongated. The dorsal surface is greenish to brown with papillae that are brown and pink in colour. The ventral surface is white to light grey and sometimes pink, with 8 to 12 rows of podia (Fig. 4.17). Twenty-five short and stout tentacles surround the ventral mouth. The anus has five anal teeth. The ossicles were found in the shape of rods in the tentacles.

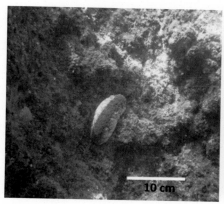 

FIGURE 4.17

The sea cucumber *Actinopyga mauritiana*.

Credit: Ahmed, M., 2009. Morphological, Ecological and Molecular Examination of the Sea Cucumber Species along the Red Sea Coast of Egypt and Gulf of Aqaba. The University of Hull.

In the dorsal body wall, the tentacles were found in the form of rods and rosettes. The ventral body wall has grains and rods (Ahmed, 2009).

The sea cucumber *A. mauritiana* was recorded in rocky and dead coral habitats in the Red Sea of Saudi Arabia (Hasan, 2009). Also, it was reported from the Red Sea coast of Egypt and the Gulf of Aqaba (Lawrence et al., 2010) in seagrass, sandy bottoms and corals (Ahmed, 2009).

The sea cucumber *A. mauritiana* from the Red Sea was recorded with a mean length of 36 ± 4.97 cm, mean width of 9.67 ± 1.53 cm, mean weight of 800 ± 180.28 g and mean thickness of the body wall 1.5 ± 0.5 cm (Eissa et al., 2017). Additionally, the gonadal development of this species was examined in the Red Sea (Gabr et al., 2005). Most mature individuals were observed during March and August. The maturity stages were described in four stages as follows: premature or recovery, maturing, mature and post-spawning (Gabr et al., 2005).

4.2.2.6 Holothuria atra *(Jaeger, 1833)*

The sea cucumber *H. atra* is often found covered by sand. The body is cylindrical, and the dorsal surface is relatively black. Twenty long and black tentacles surround the ventral mouth (Fig. 4.18). The ossicles were identified in the form of rods in the tentacles. Tables and rosettes ossicles were found in the ventral and dorsal body wall (Ahmed, 2009).

The sea cucumber *H. atra* was recorded with high density in sandy habitats with a range of 4.3 individuals $100 \, \text{m}^{-2}$ and 8.4 individuals $100 \, \text{m}^{-2}$ in the Red Sea of Saudi Arabia. The lowest densities of *H. atra* were recorded in rocky areas and dead coral habitats (Hasan, 2009). This species was found in Egypt inhabiting sandy grounds, seagrass and coral reefs (Ahmed, 2009).

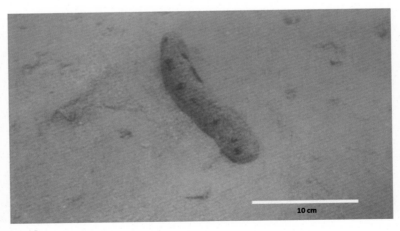

FIGURE 4.18

The sea cucumber *Holothuria atra*.

Credit: Ahmed, M., 2009. Morphological, Ecological and Molecular Examination of the Sea Cucumber Species along the Red Sea Coast of Egypt and Gulf of Aqaba. The University of Hull.

The sea cucumber *H. atra* from the Red Sea coast of Egypt was studied for its reproductive biology. The size at first maturity for males was 16.5 cm and for females 15.5 cm. Four maturity stages were detected: immature, mature, ripe and spent; the ripe stage was highly abundant from June to December for females and in June and November for males (Abdel-Razek et al., 2005). From the Red Sea coast of Egypt, the sea cucumber *H. atra* collected in 2015 from Hurghada shores. The mean length was 23.3 ± 7.6 cm, the mean width was 7.5 ± 3.1 cm and the mean weight was 368.67 ± 199.6 g (Eissa et al., 2017).

4.2.2.7 Other species

From the Red Sea coast of Egypt, the sea cucumber *H. leucospilota* was found with a mean length of 35 ± 5 cm, mean width of 4 ± 1 cm, mean live weight of 316.6 ± 79.29 g and mean body wall thickness of 0.73 ± 0.25 cm (Eissa et al., 2017). This species is found hiding the posterior body at sandy-muddy grounds with coral patches or rubble, seagrass beds, flats of the reef, shallow lagoons and black reefs (Ahmed, 2009) (Table 4.2). This species has an elongated body with black colour. Twenty black or brown tentacles surround the ventral mouth. The ossicles were found in the form of tables and buttons in the body wall (Fig. 4.19) (Ahmed, 2009). *H. leucospilota* sea cucumber positively influences the benthos life and sediment of the Red Sea in Egypt (Dar, 2004). Furthermore, Dabbagh et al. (2011) reported the induction spawning of *H. leucospilota* using heat shock or water pressure.

The sea cucumber *H. edulis* was found in the Red Sea coast of Egypt with a mean length of 23 ± 7.75 cm, mean width of 5.75 ± 1.7 cm, mean live weight of 224.3 ± 123.36 g and mean body wall thickness of 0.73 ± 0.25 cm (Eissa et al., 2017). It is found in Egypt at a depth of 30 m in shallow lagoons and among

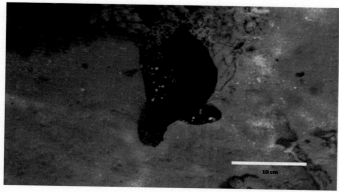

FIGURE 4.19

The sea cucumber *Holothuria leucospilota*.

Credit: Ahmed, M., 2009. Morphological, Ecological and Molecular Examination of the Sea Cucumber Species along the Red Sea Coast of Egypt and Gulf of Aqaba. The University of Hull.

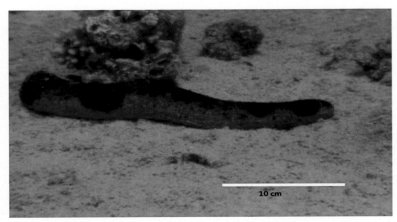

FIGURE 4.20

The sea cucumber *Holothuria edulis.*

Credit: Ahmed, M., 2009. Morphological, Ecological and Molecular Examination of the Sea Cucumber Species along the Red Sea Coast of Egypt and Gulf of Aqaba. The University of Hull.

sheltered reefs (Ahmed, 2009). It has soft and thick body wall and is roughly cylindrical. The dorsal body is black, and the ventral body is pink. Twenty grey tentacles surround the ventral mouth. The ossicles were found in the form of rods in the tentacles and in the form of tables and buttonlike rosettes in the body wall. They are similar in dorsal and ventral sides of the body wall (Fig. 4.20) (Ahmed, 2009).

The sea cucumber *Bohadschia marmorata* was found in the Red Sea coast of Egypt with a mean thickness of 1.2 ± 0.65 cm. The recorded mean length of the body was 15.83 ± 1.8 cm, the mean width was 7.6 ± 2.1 cm and the mean weight was 264.5 ± 106.35 g (Eissa et al., 2017). It inhabits coastal lagoons and inner reef gaps. It is found buried in sandy mud bottoms at depths up to 20 m (Ahmed, 2009). It has a cylindrical body in cross section with light yellowish colour ventrally, brownish dorsally, spotted by numerous small brown spots (Fig. 4.21) (Ahmed, 2009). Twenty small light brown tentacles surround the ventral mouth. The Cuvierian tubules are quickly ejected when touched. The ossicles were found in the form of spiny rods in the tentacles, grains in the ventral body wall and rosettes and grains in the dorsal body wall. In 2011, collected from the Red Sea of Egypt, namely *B. marmorata* play a crucial role in sediment nutrient recycling (Dar, 2004).

Moreover, the sea cucumber *B. vitiensis* from the Red Sea coast is observed to spawn in the summer. The size at first maturity was 24.5 cm for males and 26.1 cm for females. The maximum value observed for gonads was in June and July (Omar et al., 2013). Furthermore, *B. vitiensis* was found in the sandy bottom, seagrass beds and in dead and live corals (Fig. 4.22) (Abdel Razek et al., 2006; Hellal et al., 2007; Omar et al., 2013) (Table 4.2).

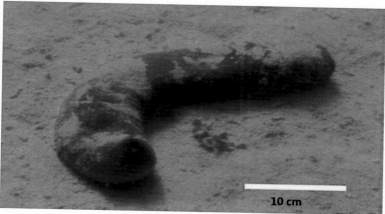

FIGURE 4.21

The sea cucumber *Bohadschia marmorata*.

Credit: Ahmed, M., 2009. Morphological, Ecological and Molecular Examination of the Sea Cucumber Species along the Red Sea Coast of Egypt and Gulf of Aqaba. The University of Hull.

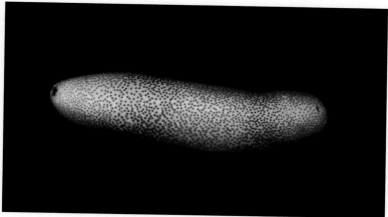

FIGURE 4.22

The sea cucumber *Bohadschia vitiensis*.

Credit: By François Michonneau via Wikimedia commons. Licensed under CC BY 3.0.

4.3 Sea cucumbers aquaculture development

4.3.1 Asexual reproduction

The asexual reproduction by fission was studied in *H. arenicola* from the Mediterranean Sea coast of Egypt. A rubber band was placed around the middle portion of the body (the upper 45%). The rubber band triggers the constriction and stretching of the body until separation within 1—5 d in the sea cucumber. The optimum temperature for fission was 30°C, and the survival rate of the posterior part was higher than of the anterior part (Abdel Razek et al., 2007). Also, the sea cucumbers *H. tubulosa* and *H. poli* from Italian waters were induced for artificial fission. The survival rate of new sea cucumbers was 85% in *H. tubulosa* and 75% in *H. poli*. Evisceration was frequently observed in the tested animals, which accounted for 50% in *H. tubulosa* and 80% in *H. poli*. Also, *H. poli* showed a longer time of regeneration (Toscano and Cirino, 2018). Furthermore, a rubber band was fitted around the mid-body of the sea cucumber *A. mauritiana* from the Red Sea, and the individuals were isolated after fission. Smaller-sized individuals (5—15 cm) were able to regenerate at a higher rate than the larger-sized individuals in around 60—80 d (Gabr et al., 2005).

4.3.2 Sexual reproduction

For the sea cucumber *H. tubulosa*, its adaptation to aquaculture was reported. The broodstock can be induced to spawn successfully, and the larvae showed a good performance in hatchery rearing. The specimens collected from the Tyrrhenian Sea between July and October were maintained in tanks with a temperature of 23—24°C. Combined thermal stimulation and thermal shock method successfully induced the sea cucumber to spawn, and the larvae reached the juvenile stage in 27 days (Rakaj et al., 2018). Günay et al. (2018) reported that the optimal stocking density for the highest hatching and survival rates of the fertilised eggs of *H. tubulosa* is 1—5 eggs/ ml (Günay et al., 2018). Also, for the juvenile (40.3 ± 3.34 g), a density of 6 ind./m^2 is recommended for stocking juvenile *H. tubulosa* under rearing conditions rather than 15 ind./m^2 and 30 ind./m^2 (Tolon et al., 2017b). Under aquaculture conditions, the highest growth performance for the juvenile was found in providing 1 mm sediment compared with 3 and 7 mm sediment (Tolon et al., 2015). Also, the best growth performance of the *H. tubulosa* larvae (average: 20.48 ± 2.33 g) was found at 25°C among other treatments at 15, 20 and 30°C (Günay et al., 2015).

The spawning induction of the sea cucumber *H. arenicola* from the Mediterranean coast of Egypt was tested in the laboratory. The injection of perivisceral coelomic fluid was the most effective method to induce spawning even in sea cucumbers that were collected before the spawning season. Also, the stripping method was useful during the spawning season, June and July. Elevating the water temperature by 5°C can also induce spawning. However, the drying method or algae path was not useful to induce spawning (Razek et al., 2012).

The sea cucumber *H. poli* from the Tyrrhenian Sea (western coast of Italy, Mediterranean Sea) was artificially produced in the laboratory. Specimens collected from July to September were maintained at a water temperature of 24°C and successfully induced to spawn in July to September by thermal shock (see chapter 6 for further illustration on spawning methods). The larvae performed satisfactorily in the hatchery with feeding concentrations of 20,000—40,000 cells/mL microalgae (Rakaj et al., 2019).

4.4 Sea cucumber utilisation

4.4.1 Ecological values

4.4.1.1 Sea cucumber activities improve sediment characteristics

Sea cucumbers can be used successfully for the improvement of coastal regions in Egypt with its associated fauna (Kazzaz et al., 2019; Shabana et al., 2018). Also, the isotopic analysis revealed that *H. poli* and *H. tubulosa* have a selection/preferential assimilation of seagrass detritus, which highlights their role in the organic matter cycling derived from the seagrass meadows in Mediterranean coastal ecosystems (Boncagni et al., 2019).

The ecological role of three holothurian species was evaluated in the Red Sea of Egypt through analysing their gut content (Dar, 2004). Six sea cucumber species were collected from four sites along the Red Sea coast of Egypt (Dar, 2004). Holothurians consume a large amount of sediment during the feeding period, leading to nutrient recycling, and refresh the sediment characteristics (Dar, 2004). The annual amount of the reworked sediment was 45.78 kg/yr, 28.72 kg/yr, 21.23 kg/yr, 13.10 kg/yr, 23.64 kg/yr and 26.32 kg/yr for *B. marmorata*, *H. atra*, *H. leucospilota*, *H. scabra*, *H. edulis* and *A. mauritiana*, respectively (Dar, 2004).

In 2006, the ecological role of three sea cucumber species was investigated from three locations in the Red Sea of Egypt. The authors concluded that holothurians could cause significant changes in the composition of the seafloor sediment. Sediment reworking of the three species (i.e. *B. vitiensis*, *H. atra* and *H. hawaiiensis*) was varied according to local conditions and individual size, numbers and sexual maturity stage (Dar and Ahmad, 2006). Also, the intensive reworking of holothurians started in the late spring until the end of autumn (Dar and Ahmad, 2006). Furthermore, the reworking of the sea cucumbers increased during the reproduction period (maturation, spawning and post-spawning). The most significant reworking operation occurred from late spring to the end of the summer (Dar and Ahmad, 2006). The reworking of the sea cucumber *B. vitiensis* was 36.71 kg/yr ind. in summer and 44.73 kg/yr ind. in August, for the sea cucumber *H. atra* was 64.36 kg/yr ind. in July and 27.77 kg/yr ind. in summer and for the sea cucumber *H. hawaiiensis* was 54.88 kg/yr ind. in December and 51.34 kg/yr ind. in autumn (Dar and Ahmad, 2006).

In protected aquaculture ponds in Hurghada, El Gouna Resort, Red Sea of Egypt, the influence of the sea cucumber *H. atra* on the sediment was evaluated through analysing the foregut and hindgut of the intestine (Shabana et al., 2018). Ammonification rate and nitrification rate of the ingested sediment in the foregut were higher than in hindgut contents. The presence of the sea cucumber increases the redox potential of sediment and keeps its characteristics in the oxidised form, which enhances biological activities (Shabana et al., 2018). Furthermore, the total organic matter of the sediment decreased in the presence of the sea cucumber by 16.4%, 5.34%, 3.02% and 3.61% of the high-density sea cucumber (108 ind. 30 m^{-2}), medium density (54 ind. 30 m^{-2}), low density (18 ind. 30 m^{-2}) and control (empty of the sea cucumber), respectively (Shabana et al., 2018).

Furthermore, sea cucumbers stocked in fences improved the sediment characteristics. Sea cucumbers were stocked in fences with a density of 60 individuals per fence of 9 m^2, ten individuals per fence of 9 m^2 and control (without sea cucumbers) (Hanafy, 2011). The sediment in the fences populated with sea cucumbers contained lower concentrations of organic carbon, bacteria and chlorophyll A than of the control, whereas the concentrations of ammonia, nitrite and nitrate and phosphorus were higher in cages populated with *A. mauritiana* than of the control cages (Hanafy, 2011). Therefore, sea cucumbers are essential for maintaining and improving the quality of the sediment and nutrient recycling (Hanafy, 2011).

Increasing aquaculture activities can cause unfavourable effects on the water quality parameters, sediment characteristics and benthic organisms in the Aegean Sea (Yucel-Gier et al., 2007). The sea cucumber *H. tubulosa* collected from Izmir Bay of the Aegean Sea can be used for the improvement of coastal regions as it can improve the water and sediment with the associated fauna, especially in aquaculture areas (Işgören-Emiroǧ;lu and Günay, 2007). Also, there was a similar observation with the sea cucumber *H. sanctori* (Felaco Durán, 2014). Furthermore, the sea cucumber *H. tubulosa* was successfully bred under net cages of fish without additional feed cost in the Mersin Bay northeast of the Mediterranean Sea (Tolon et al., 2017a). However, the sea cucumbers should be inserted in modulated box cages to prevent them from escaping and avoid predators (Tolon et al., 2017b). In the Eastern Mediterranean Sea, the sea cucumber *Actinopyga bannwarthi* was able to obtain energy and nutrients from the gilthead seabream wastes, emphasising their potential as integrated multitrophic aquaculture (IMTA) species (Israel et al., 2019). Moreover, the sea cucumber *H. tubulosa* reduces the total organic load of the fish farm biodeposits (seabream *Sparus aurata* and seabass *Dicentrarchus labrax*) in the Eastern Mediterranean Sea, Pagasitikos Gulf (Neofitou et al., 2019).

4.4.1.2 Sea cucumbers improve the surrounding environment
A significant effect of the sea cucumber *H. atra* on the pH was observed as well as the number of the culturable bacteria (total viable bacteria, *Vibrio* sp., total coliform, *Salmonella* and *Shigella*) (Kazzaz et al., 2019). Also, the existence of sea cucumbers can positively influence water quality (Dar and Ahmad, 2006). Furthermore, 43 *Carapus* sp. (a genus of pearlfish) were found in 1588 *H. poli*, 26 in 1288 *H. tubulosa* and 8 in 498 *H. mammata* (Aydin and Erkan, 2015), which highlights the role of these species in enhancing ecosystem biodiversity. Also, the pearlfish *Carapus acus* was found to live in association with the sea cucumber *Parastichopus regalis* (González-Wangüemert et al., 2014). Moreover, *H. tubulosa* is considered to be bioindicator species in determining Ni trace metal in sediment (Turk Culha et al., 2016).

Furthermore, the sea cucumber collected from the Adriatic Sea can be used as a partial substitute of fish meal in seabream nutrition (Piccinno et al., 2013), which highlights the utilisation of the sea cucumber as an alternative protein source for an extensively cultured fish species in the Mediterranean.

4.4.2 Nutritional and medicinal values

The nutritional and medicinal benefits of sea cucumbers have received significant attention from researchers in a try to convince the consumers with their benefits. With the importance of the economic development in sea cucumber aquaculture, together with the observed low conservation values for sea cucumbers, the linkage of the bioactive values of sea cucumber species to conservation is essential for the maintenance and sustainable exploitation of sea cucumber species (Lawrence et al., 2010).

4.4.2.1 Nutritional values

The sea cucumber *A. mauritiana*, *H. scabra*, *B. marmorata* and *H. leucospilota* collected from the Red Sea coast of Egypt have high nutritional values, and their bioactive compounds are useful in tissue regeneration and inflammatory disease (Omran, 2013). Also, *A. mauritiana* has the highest protein content (48.27%), and *H. scabra* is rich in omega-6 (Omran, 2013). Furthermore, the sea cucumbers *H. scabra*, *A. echinites* and *H. atra* collected from the Sudanese Red Sea have a high protein content, and minimal content of fats, which encourages the utilisation of the sea cucumber as a food commodity (Ibrahim et al., 2015). The sea cucumber *H. tubulosa* collected from the Aegean Sea is rich in amino acid and fatty acid content. The chemical composition for the sea cucumber was 84.04% moisture, 10.33% protein, 0.20% fat, and 6.63% ash (Öztürk and Gündüz, 2018). Likewise, the sea cucumber *Holothuria froskali* collected from Izmir Bay of the Aegean Sea has high protein content. It can be utilised as a protein source, and it has a high rate of long-chain fatty acids and rich in eicosapentaenoic acid (EPA) and docosahexaenoic acid (Bilgin and İZCİ, 2015).

Many studies have highlighted the sea cucumber as a source of polyunsaturated fatty acids (PUFA). *H. tubulosa* and *H. poli* collected in a coastal area of the Adriatic Sea contained very low lipid and were rich in PUFA. EPA and eicosatrienoic acid were the highest abundant among PUFA in both species (Sicuro et al., 2012). Additionally, glycine, glutamic acid and aspartic acid were the most abundant amino acids among the total amino acids observed. Furthermore, the heavy metals concentrations in the analysed species were under the maximum limits for the Italian legislation (Sicuro et al., 2012). Also, the sea cucumber *P. regalis* from San Pedro del Pinatar (Murcia, Spain) at the shores of Mediterranean Sea has a high moisture content, moderate protein and low lipid levels. Glutamic acid, arginine and tyrosine were the most abundant amino acids observed. Also, PUFA, especially arachidonic acid, dominated the fatty acid profile. Iron, sodium, calcium and zinc were the most abundant minerals (Roggatz et al., 2018). Likewise, three species of sea cucumbers harvested from the Aegean Sea are rich in PUFA (Aydin et al., 2011). Also, the sea cucumber *H. sanctori* collected from the Mediterranea Sea coast of Turkey has a good range of PUFA/saturated fatty acid and n3:n6 ratios were found to be 1.05 and 1.87. However, it has a low amount of minerals and fatty acids when compared with other commercial species (Göçer et al., 2018).

The chemical composition of the sea cucumber species varies according to the season, species and region. Özer et al. (2005) collected the sea cucumbers *H. scabra* from the Marmara Sea during spring and summer for chemical composition analysis. The sea cucumber was harvested in April, May, June, July and August. Two methods were used to process the sea cucumber. The first method was conducted by cutting the anus and removing the viscera. The second method was performed by cutting along the body and removing the viscera. The second method resulted in low protein content in July (Özer et al., 2005). Also, the biochemical composition of the sea cucumber *H. tubulosa* from three stations (Gelibolu, Umurbey and Dardanos) in Dardanelles Strait, Turkey, varies by season and location. The protein content was 52.48% in March at Gelibolu, 52.81% in May at Umurbey and 56.93% at Dardanos. Also, lipid was 1,.70% in February at Gelibolu; in March was 1,.77% at Umurbey and in September was 1,.42% at Dardanos. Furthermore, ash was 43.51 % in November at Gelibolu; in September was 44.22% at Umurbey and in December was 44.98% at Dardanos (Culha et al., 2017). Furthermore, the approximate composition of the sea cucumbers *H. mammata*, *H. poli* and *H. tubulosa* ranged between 81.24% and 85.24%, 7.88% and 8.82%, 0.09% and 0.18% and 5.13% and 7.85% for moisture, protein, fat and ash contents, respectively (Aydin et al., 2011).

The nutritional composition of the sea cucumber *H. tubulosa* varies according to the drying methods. Microwave drying was found more suitable than hot air drying and freeze-drying to preserve the amino acid and fatty acid content within a sufficient time (Öztürk and Gündüz, 2018). However, freeze-drying method helped to preserve the protein and the fat content as their values were high in samples dried via freeze-drying method (Öztürk and Gündüz, 2018). Furthermore, the oven drying method was recommended for healthy consumption because ambient drying is conducted under uncontrolled atmospheric conditions (Bilgin and Öztürk Tanrikulu, 2018). However, the nutritional composition of sea cucumbers was higher when the sea cucumber dried via ambient temperature (23±2°C) comparing with oven drying (45±1°C). Also, the proximate composition was calculated after sun drying of 8 d for sea cucumber *H. tubulosa* collected from the Aegean Sea region. This drying period did not trigger oxidation in the dried product, and no criteria of quality deterioration reached the maximum limit for the acceptability of the product (Çakh et al., 2004). Likewise, sun drying the sea cucumber using solar dryer was found to be effective than the traditional method using sun drying. The sun drying method consumes 1−2 d and reduces the loss or damage that might result from insects, birds, rodents and adverse climatic conditions (Vaipulu, 2009).

Various bioactive compounds were isolated from sea cucumbers. Monosulphated biosides, holothurins B_2, B_3, B_4 and the holothurins A and B (Figs. 4.23 and 4.24) were isolated from the sea cucumbers *H. poli*, *H. tubulosa* and *Holothuria* sp. collected in Capo Miseno Bay in the Mediterranean Sea, Italy (Silchenko et al., 2005). Three monosulphated biosides (i.e. holothurians B_2, B_3 and B_4) were isolated from the sea cucumber *H. poli*. Triterpene glycosides belonging to holothurin A and B groups were found in *H. tubulosa*. In contrast, only one type of glycoside, i.e. holothurin A (4), was isolated from *Holothuria* sp. (Silchenko et al., 2005). Also, the

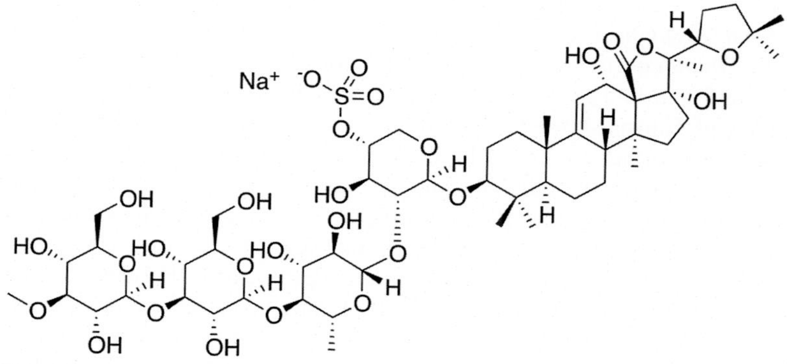

FIGURE 4.23

The chemical structure of holothurin A compound.

Credit: Public domain.

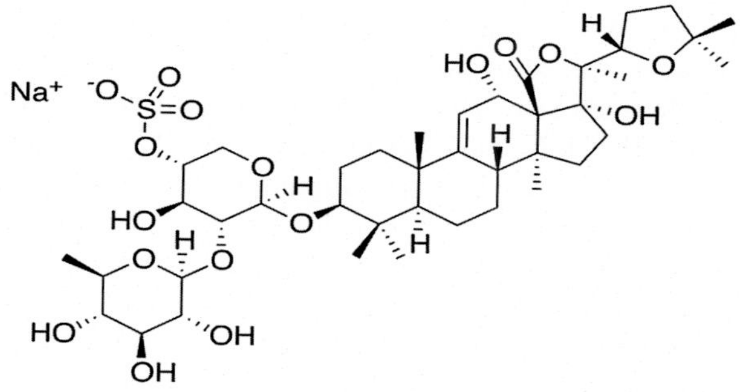

FIGURE 4.24

The chemical structure of holothurin B compound.

Credit: Public domain.

sea cucumber *H. forskali* collected from Banyuls-sur-Mer (southern France) at a depth of 20–30 m contains a high amount of saponins. Twelve saponins have been detected in the body wall and 26 in the Cuvierian tubules (Van Dyck et al., 2009). Furthermore, 18 different saponins were isolated from the sea cucumber *H. sanctori* collected from the Algerian coast. Higher diversity of saponins is found in the body wall (12) than in the Cuvierian tubules (8) (Caulier et al., 2016).

4.4.2.2 Anti-bacterial properties

Several studies have reported the potential of using the coelomic fluid, the body wall, the internal organs and the extracted bioactive compound of sea cucumbers as an anti-bacterial agent. The coelomic fluid and flesh of the sea cucumber *H. scabra*, *H. leucospilota* and *H. atra*, which were collected from the Red Sea coast of Egypt, are promising natural antibiotic. The crude extract of the coelomic fluid and the body wall from the three sea cucumber species showed anti-bacterial effect against *Pseudomonas aeruginosa* ATCC 8739, *Staphylococcus aureus* ATCC 6538, *Escherichia coli*, *Streptococcus faecalis* and *Vibrio damsela* (Ibrahim, 2012). Also, the coelomic fluid and body wall extract of the sea cucumber *H. tubulosa* collected from the bay of Tunisia showed anti-microbial and anti-fungal activity. The highest growth inhibitory effect was found in the body wall extracts. The most potent anti-bacterial extracts were the ethanol extracts. The ethanol and butanol extracts of *H. tubulosa* revealed high activity against both Gram-negative and Gram-positive bacteria. Additionally, extensive activities against yeast (*Candida ciferrii*) and fungi (*Geotrichum penicillatum*) were observed (Sellem et al., 2017). Furthermore, the sea cucumber *H. tubulosa* collected from the Gulf of Palermo (southern Italy) was tested for its anti-microbial agents in the coelomic fluid. The immune mediators in *H. tubulosa* are a source of anti-microbial peptides for the development of useful anti-microbial agents against biofilm bacterial communities that are often intrinsically resistant to conventional antibiotics (Schillaci et al., 2013).

The body wall extracts of the sea cucumber *H. atra* and *H. nobilis* from the Red Sea of Egypt showed anti-bacterial activity against *S. aureus* (ATCC6538), *P. aeruginosa* (ATCC6739) as Gram-positive and *E. coli* as Gram-negative (Soliman et al., 2016). Likewise, the body wall extract of the sea cucumber *H. tubulosa* collected from the Marmara Sea showed anti-microbial activity against *Salmonella typhimurium* (ATCC 51812), *E. coli* (ATCC 25922) and *S. aureus* (ATCC 25923), except for *Bacillus cereus* (ATCC 7464) (Bilgin and İZCİ, 2015). Moreover, the extract of the body wall, intestine and gonads of the sea cucumber *H. poli* from the Mediterranean coast of Egypt showed anti-bacterial activity against *Aeromonas hydrophila* and *Salmonella choleraesuis* (ATCC 14028) (Omran and Allam, 2013).

However, some studies reported non-anti-bacterial activity for some sea cucumber species. For instance, holothurin B compound (belongs to the class of compounds known as saponins) was isolated from the sea cucumber *H. arenicola* collected from the Mediterranean coast of Egypt. However, it showed no anti-bacterial activity against *Vibrio fluvialis*, *S. aureus*, *E. coli*, *S. faecalis* and *P. aeruginosa* (El Nemr et al., 2012). Likewise, 11 sea cucumber species collected from the Red Sea of Egypt, namely, *H. leucospilota*, *H. nobilis*, *H. atra*, *H. fuscogilva*, *S. hermanni*, *A. mauritiana*, *Actinopyga crassa*, *B. vitiensis*, *Bohadschia tenuissima*, *B. cousteaui* and *P. graeffei*, and their body wall extract showed no anti-bacterial activity; however, all extracts from the 11 sea cucumbers showed activity against *Candida*, *Leishmania* and LoVo cell lines (Lawrence et al., 2010).

4.4.2.3 Anti-fungal properties

Several sea cucumber species are reported to have anti-fungal activity. The body wall, the intestine and the gonads extract from the sea cucumber *H. poli*, which was collected from the Mediterranean coast of Egypt, showed an anti-fungal effect, particularly against *Aspergillus niger*, *Scloretium* sp., *Candida albicans*, *Aspergillus flavus* and *Malassezia furfur* (Omran, 2006; Omran and Allam, 2013). Also, both the aqueous and the methanolic extracts of the sea cucumber *H. poli* collected from

Tunisia coast were found to have significant anti-fungal activity. This activity was observed from the semi-purified fractions (F5, F6) of both extracts in a concentration-related manner against *Aspergillus fumigatus* higher than *Trichophyton rubrum*. However, no activity was observed against strains of *C. albicans* (Ismail et al., 2008). Furthermore, 21 lanostane-type non-sulphated triterpene glycosides were isolated from the methanol/methylene chloride extract of the body walls of *B. cousteaui* from the Red Sea coast of Egypt, and most of them have an anti-fungal influence on *C. albicans* (Elbandy et al., 2014). Moreover, the crude extract of the sea cucumber *Bohadschia graeffei* from the Red Sea coast of Egypt has a potent anti-fungal activity on the pathogenic fungal species *C. albicans* (Khattab et al., 2008).

4.4.2.4 Other properties

Sea cucumbers were explored for their potential as anti-cancer, anti-parasitic, anti-coagulant, anti-fouling, anti-oxidant and anti-inflammatory agents. The bioactive compounds of *H. poli* from the Mediterranean coast of Egypt have shown anti-tumour effect (Omran and Khedr, 2015). Specifically, the glycoside (non-sulphated hexaosides called bivittoside) extracted from the body wall is effective against colon adenocarcinoma tumour cell line and breast adenocarcinoma tumour cell lines (Omran and Khedr, 2015). Also, the saponin extracted from the body wall of the sea cucumber *H. arenicola* collected from Mediterranean Sea coast of Egypt showed anti-tumour activity against Ehrlich Ascites Carcinoma tumour model in female Swiss albino mice (Mohamed et al., 2019). Furthermore, saponin compounds isolated from the sea cucumber *P. graeffei* collected from the Red Sea coast of Egypt may be used as anti-cancer, anti-fungal and anti-leishmanial agents (Khattab et al., 2018).

The body wall ethanol extract of the sea cucumber *H. poli* and *A. mauritiana* and the crude extract of the Cuvierian tubes of the sea cucumber *B. vitiensis* have anti-schistosomal influence (Mona et al., 2012). Also, the isolated echinosides A and B from the sea cucumbers *A. echinites* and *H. poli* possess potential in vitro activity against *S. mansoni* adult worms (Fig. 4.25) (Melek et al., 2012). Moreover, the ethanolic extract of the sea cucumber *H. atra* collected from Northern and Southern Corniche of Jeddah contains promising constituents as insecticides against the larvae of the mosquito vector of dengue fever (Mahyoub et al., 2016).

The fucosylated chondroitin sulphate (Fig. 4.26) isolated from the sea cucumber *H. poli* body wall collected from the Mediterranean Sea exhibited a high anti-coagulant effect (Ben Mansour et al., 2017). Also, the body wall extracts of the sea cucumber *H. atra* and *H. nobilis* from the Red Sea of Egypt showed anti-fouling activity (Soliman et al., 2016). Moreover, the body wall extract of the sea cucumber *H. arenicola* from the Mediterranean coast of Egypt has proven its efficiency against renal injury induced by bile duct ligation in male albino rats (Fahmy and Mohamed, 2015).

Sea cucumbers were investigated for their remarkable anti-oxidant activity. The body wall extract of the sea cucumber *H. atra* reduces doxorubicin-induced

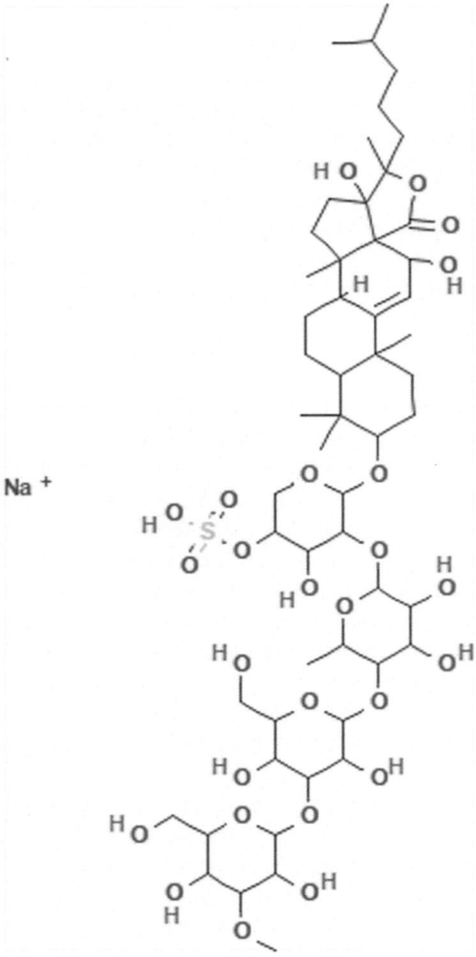

FIGURE 4.25

A two-dimensional representaion of the echinoside A.

Credit: PubChem. PubChem Identifier: CID 156831 https://pubchem.ncbi.nlm.nih.gov/compound/Echinoside-A.

cardiotoxicity via their anti-oxidant and anti-apoptotic activities, which might be useful in the treatment of human patients under doxorubicin chemotherapy (Ibrahim et al., 2017). Also, a remarkable hepatoprotective effect on the injury liver of rat models has been shown when they treated by the body wall mixed extract from the sea cucumber *H. atra* from the Red Sea of Egypt (Esmat et al., 2013). Furthermore, the saponin isolated from the sea cucumber *Holothuria thomasi*, which was collected from the Red Sea coast of Egypt, has the potential of anti-oxidant and anti-inflammatory activities and may be useful in improving dyslipidemia,

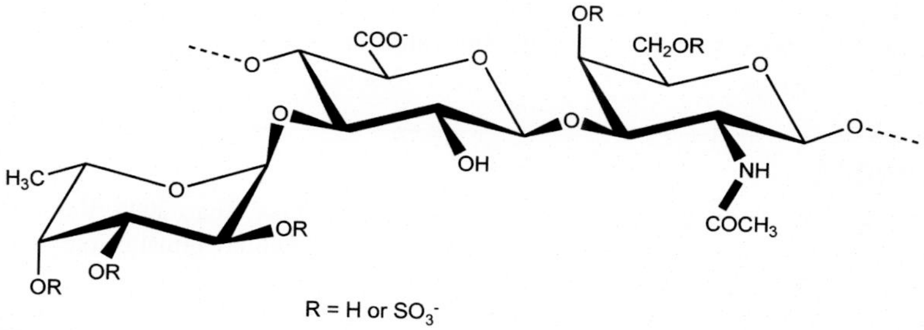

FIGURE 4.26

The chemical structure of fucosylated chondroitin sulphate.

Credit: Pomin, V.H., 2014. Holothurian fucosylated chondroitin sulfate. Mar. Drugs, 12(1), 232–254. https://doi.org/10.3390/md12010232.

controlling glycaemic status and diminishing diabetic cardiovascular complications (El Barky et al., 2016). In addition to its anti-bacterial activity, the sea cucumber *H. tubulosa* can be considered as a natural anti-oxidant agent (Künili and Çolakoğlu, 2018). Moreover, the aqueous extract of the body wall has a positive influence against isoproterenol-induced cardiac injury in vivo and in vitro of the rats (Esmat et al., 2012).

4.5 Conclusion remarks

Despite the growing interest in sea cucumber aquaculture in the Mediterranean and the Red Seas, no studies concerned with the commercial production. Trials of sea cucumber aquaculture for both sexual and asexual production have been carried mainly on the sea cucumbers *H. tubulosa* and *H. poli* and *H. arenicola*. Developing sea cucumbers aquaculture will not only provide a source of high income, given their economic value (although their role in the IMTA needs to be further explored in the region), but also will introduce many ecosystem services (given their observed rule in the ecosystem) and will present a wide range of nutritional and medicinal benefits to the consumers (given their high nutritional and medicinal values). The review of habitat shows how it influences the distribution of sea cucumber species. Reproductive biology studies, in both the field and the laboratory, demonstrate how environmental conditions affect spawning. Therefore, this knowledge is essential for successful sea cucumber aquaculture.

Sea cucumbers in the Mediterranean are increasingly considered as an important source of food and medicine. Literature reported that sea cucumbers have high nutritive value and their bioactive compounds have multiple biological activities, such as anti-bacterial and anti-fungal. These values vary according to species, season, region and processing method. The linkage of the bioactive values of sea cucumber species to conservation is essential for the maintenance and sustainable exploitation of sea cucumber species. Overall, sea cucumber can be explored as a potential source of high-value components to develop valuable functional food.

References

Abdel-Razek, F.A., Abdel-Rahmen, S.H., El-Shumy, N.a., Omar, H.a., 2005. Reproductive biology of the tropical sea cucumber *Holothuria atra* (Echinodermata:Holothuroidea) in the Red Sea coast of Egypt. Egypt. J. Aquat. Res. 31, 383—402.

Abdel Razek, F.A., Abdel Rahman, S.H., Mona, M.H., El-Gamal, M.M., Moussa, R.M., 2007. An observation on the effect of environmental conditions on induced fission of the Mediterranean sand sea cucumber, *Holothuria arenicola* in Egypt. Beche-de-mer Inf. Bull. 26, 33—34.

Abdel Razek, F.A., El-Shimy, N.A., Abdel Rahman, S.H., Omar, H.A., 2006. Ecological observations on the abundance, distribution of holothuroids (Echinodermata—Holothuroidea) in the Red Sea coast, Egypt. Egypt J. Aquat. Res. 32, 346—361.

Ahmed, M., 2009. Morphological, Ecological and Molecular Examination of the Sea Cucumber Species along the Red Sea Coast of Egypt and Gulf of Aqaba. The University of Hull.

Ahmed, M.I., Lawrence, A.J., 2007. The status of commercial sea cucumbers from Egypt's northern Red Sea Coast. SPC Beche-de-mer Inf. Bull. 26, 14—18.

Aydin, M., 2008. The commercial sea cucumber fishery in Turkey. SPC Beche de Mer Information Bulletin, 28, pp. 40—41.

Aydin, M., 2016. Sea cucumber (Holothuroidea) species of Turkey Türkiye Denizlerindeki deniz hıyarı (Holothuroidea) Türleri. Turkish J. Maritime Mar. Sci. 2 (1), 49—58.

Aydin, M., 2019. Biometry, density and the biomass of the commercial sea cucumber population of the Aegean Sea. Turkish J. Fish. Aqua. Sci. 19 (6), 463—474.

Aydin, M., Erkan, S., 2015. Identification and some biological characteristics of commercial sea cucumber in the Turkey coast waters. Int. J. Fish. Aquat. Stud. 3, 260—265.

Aydin, M., Sevgili, H., Tufan, B., Emre, Y., Köse, S., 2011. Proximate composition and fatty acid profile of three different fresh and dried commercial sea cucumbers from Turkey. Int. J. Food Sci. Technol. 46, 500—508. https://doi.org/10.1111/j.1365-2621.2010.02512.x.

Bailey, G., 2010. The Red Sea, coastal landscapes, and hominin dispersals. In: Vertebrate Paleobiology and Paleoanthropology. Springer, pp. 15—37. https://doi.org/10.1007/978-90-481-2719-1_2.

Bardanis, E., Batjakas, I., 2018. November. Reproductive cycle of *Holothuria poli* in Gera Gulf, Lesvos Island, Greece. In: 3rd Int. Congr. Appl. Ichthyol. Aquat. Environ, pp. 8—11.

Ben Mansour, M., Balti, R., Ollivier, V., Ben Jannet, H., Chaubet, F., Maaroufi, R.M., 2017. Characterization and anticoagulant activity of a fucosylated chondroitin sulfate with unusually procoagulant effect from sea cucumber. Carbohydr. Polym. 174, 760—771. https://doi.org/10.1016/j.carbpol.2017.06.128.

Bilgin, Ş., İZCİ, L., 2015. The effects of drying & boiling process on nutritional components of *Holothuria forskali* (Delle Chiaje, 1823). J. Food Health Sci. 2, 1—8. https://doi.org/10.3153/jfhs16001.

Bilgin, Ş., Öztürk Tanrikulu, H., 2018. The changes in chemical composition of *Holothuria tubulosa* (Gmelin, 1788) with ambient-drying and oven-drying methods. Food Sci. Nutr. 6, 1456—1461. https://doi.org/10.1002/fsn3.703.

Boncagni, P., Rakaj, A., Fianchini, A., Vizzini, S., 2019. Preferential assimilation of seagrass detritus by two coexisting Mediterranean Sea cucumbers: *Holothuria polii* and *Holothuria tubulosa*. Estuar. Coast. Shelf Sci. 231, 106464. https://doi.org/10.1016/j.ecss.2019.106464.

Borrero-Pérez, G.H., González-Wangüemert, M., Marcos, C., Pérez-Ruzafa, A., 2011. Phylogeography of the Atlanto-Mediterranean Sea cucumber *Holothuria* (Holothuria) *mammata*: the combined effects of historical processes and current oceanographical pattern. Mol. Ecol. 20, 1964—1975. https://doi.org/10.1111/j.1365-294X.2011.05068.x.

Bulteel, P., Jangoux, M., Coulon, P., 1992. Biometry, bathymetric distribution, and reproductive cycle of the holothuroid *Holothuria tubulosa* (Echinodermata) from Mediterranean Sea grass beds. Mar. Ecol. 13, 53—62. https://doi.org/10.1111/j.1439-0485.1992.tb00339.x.

Çaklı, Ş., Cadun, A., Kişla, D., Dinçer, T., 2004. Determination of quality characteristics of *Holothuria tubulosa*, (Gmelin, 1788) in Turkish Sea (Aegean Region) depending on sun drying process step used in Turkey. J. Aquat. Food Prod. Technol. 13, 69—78. https://doi.org/10.1300/J030v13n03_07.

Caulier, G., Mezali, K., Soualili, D.L., Decroo, C., Demeyer, M., Eeckhaut, I., Gerbaux, P., Flammang, P., 2016. Chemical characterization of saponins contained in the body wall and the cuvierian tubules of the sea cucumber *Holothuria* (Platyperona) *sanctori* (Delle Chiaje, 1823). Biochem. Systemat. Ecol. 68, 119—127. https://doi.org/10.1016/j.bse.2016.06.005.

Coleman, R.G., 1974. Geologic background of the Red Sea. In: The Geology of Continental Margins. Springer Berlin Heidelberg, pp. 743—751. https://doi.org/10.1007/978-3-662-01141-6_55.

Coll, M., Piroddi, C., Steenbeek, J., Kaschner, K., Lasram, F.B.R., Aguzzi, J., Ballesteros, E., Bianchi, C.N., Corbera, J., Dailianis, T., Danovaro, R., 2010. The biodiversity of the Mediterranean Sea: estimates, patterns, and threats. PloS One 5 (8). https://doi.org/10.1371/journal.pone.0011842.

Coulon, P., Jangoux, M., 1993. Feeding rate and sediment reworking by the holothuroid *Holothuria tubulosa* (Echinodermata) in a Mediterranean seagrass bed off Ischia Island, Italy. Mar. Ecol. Prog. Ser. 92, 201—204. https://doi.org/10.3354/meps092201.

Culha, S.T., Çelik, M.Y., Karaduman, F.R., Dereli, H., Culha, M., Ozalp, H.B., Hamzacebi, S., Alparslan, M., 2017. Influence of seasonal environmental changes on the biochemical composition of sea cucumber (*Holothuria tubulosa* Gmelin, 1791) in the Dardanelles Strait. Ukr. Food J. 6, 291—301. https://doi.org/10.24263/2304-974x-2017-6-2-10.

Dabbagh, A.R., Sedaghat, M.R., Rameshi, H., Kamrani, E., 2011. Breeding and larval rearing of the sea cucumber *Holothuria leucospilota* Brandt (*Holothuria vegabunda* Selenka) from the northern Persian Gulf, Iran. SPC Beche-de-mer Inf. Bull. 31, 35—38.

Dar, M.A., 2004. Effects of anthropogenic activities in Red Sea ports on marine environment. Holothurian Role in the marine sediments reworking processes, vol. 12. Sedimentology of Egypt, pp. 173—183.

Dar, M.A., Ahmad, H.O., 2006. The feeding selectivity and ecological role of shallow water holothurians in the Red Sea. SPC Beche-de-mer Inf. Bull. 11—21.

Dereli, H., Çulha, S.T., Çulha, M., Özalp, B.H., Tekinay, A.A., 2016. Reproduction and population structure of the sea cucumber *Holothuria tubulosa* in the Dardanelles Strait, Turkey. Mediterr. Mar. Sci. 17, 47—55. https://doi.org/10.12681/mms.1360.

Despalatović, M., Grubelić, I., Šimunović, A., Antolić, B., Žuljević, A., 2004. Reproductive biology of the holothurian *Holothuria tubulosa* (Echinodermata) in the Adriatic Sea. J. Mar. Biol. Assoc. U. K. 84, 409—414. https://doi.org/10.1017/S0025315404009361h.

Drake, C.L., Girdler, R.W., 1964. A geophysical study of the Red Sea. Geophys. J. Roy. Astron. Soc. 8, 473—495. https://doi.org/10.1111/j.1365-246X.1964.tb06303.x.

Eissa, S., Omran, N., Salem, H., Kabbash, A., Kandeil, M., 2017. Surveillance study on the most common sea-cucumbers in some Egyptian coasts. Egypt. J. Exp. Biol. 13, 1. https://doi.org/10.5455/egysebz.20171113105752.

El-Ganainy, A., Hasan, M., Aquatic, M.Y.-E.J., 2006. Population structure of two endangered holothurian species from the Gulf of Aqaba, Red Sea, Egypt. Egypt. J. Aquat. Res. 32, 456–467.

El Barky, A.R., Hussein, S.A., Alm-Eldeen, A.A., Hafez, Y.A., Mohamed, T.M., 2016. Anti-diabetic activity of *Holothuria thomasi* saponin. Biomed. Pharmacother. 84, 1472–1487. https://doi.org/10.1016/j.biopha.2016.10.002.

El Nemr, A., Abdel Razek, F.A., El-Sikaily, A., Abd El-Rahman, S.H., Moussa, R.M., Taha, S.M., 2012. Isolation of holothurin B from sea cucumber *Holothuria arenicola* collected from Alexandria coast, Egypt. Blue Biotechnol. J. 1, 43.

Elbandy, M., Rho, J.R., Afifi, R., 2014. Analysis of saponins as bioactive zoochemicals from the marine functional food sea cucumber *Bohadschia cousteaui*. Eur. Food Res. Technol. 238, 937–955. https://doi.org/10.1007/s00217-014-2171-6.

El-Naggar, A.M., Ashaat, N.A., El-Belbasi, H.I., Slama, M.S., 2008. Molecular phylogeny of Egyptian sea cucumbers as predicted from 16s Mitochondrial rRNA gene sequences. World Appl. Sci. J. 5 (5), 531–542.

Esmat, A.Y., Said, M.M., Hamdy, G.M., Soliman, A.A., Khalil, S.A., 2012. In vivo and in vitro studies on the antioxidant activity of aloin compared to doxorubicin in rats. Drug Dev. Res. 73, 154–165. https://doi.org/10.1002/ddr.21006.

Esmat, A.Y., Said, M.M., Soliman, A.A., El-Masry, K.S., Badiea, E.A., 2013. Bioactive compounds, antioxidant potential, and hepatoprotective activity of sea cucumber (*Holothuria atra*) against thioacetamide intoxication in rats. Nutrition 29, 258–267. https://doi.org/10.1016/j.nut.2012.06.004.

Fahmy, S.R., Mohamed, A.S., 2015. *Holoturia arenicola* extract modulates bile duct ligation-induced oxidative stress in rat kidney. Int. J. Clin. Exp. Pathol. 8, 1649–1657.

Felaco Durán, L., 2014. Evaluation of a Multitrophic Biofiltration System with New Algae Species and the Sea Cucumber *Holothuria Sanctori* (Master thesis).

Gabr, H.R., Ahmed, A.I., Hanafy, M.H., Lawrence, A.J., Ahmed, M.I., El-Etreby, S.G., 2005. Mariculture of sea cucumber in the Red Sea -the Egyptian experience. FAO Fish. Aquac. Tech. Pap. 373–384.

Gharbi, A., Said, K., 2011. Genetic variation and population structure of *Holothuria polii* from the eastern and western Mediterranean coasts in Tunisia. J. Mar. Biol. Assoc. U. K. 91, 1599–1606. https://doi.org/10.1017/S0025315411000245.

Göçer, M., Olgunoglu, I.A., Olgunoglu, M.P., 2018. A study on fatty acid profile and some major mineral contents of sea cucumber (*Holothuria* (platyperona) *sanctori*) from Mediterranean Sea a study on fatty acid profile and some major mineral contents of sea cucumber (*Holothuria* (platyperona) *sanctor*. Food Sci. Qual. Manag. 72.

González-Wangüemert, M., Borrero-Pérez, G., 2012. A new record of *Holothuria arguinensis* colonizing the Mediterranean Sea. Mar. Biodivers. Rec. 5. https://doi.org/10.1017/S1755267212000887.

González-Wangüemert, M., Maggi, C., Valente, S., Martínez-Garrido, J., Rodrigues, N.V., 2014. *Parastichopus regalis*- the main host of *Carapus acus* in temperate waters of the Mediterranean Sea and northeastern Atlantic Ocean. SPC Beche-de-mer Inf. Bull. 38–42.

González-Wangüemert, M., Valente, S., Aydin, M., 2015. Effects of fishery protection on biometry and genetic structure of two target sea cucumber species from the Mediterranean Sea. Hydrobiologia 743, 65–74. https://doi.org/10.1007/s10750-014-2006-2.

Günay, D., Emiroğlu, D., Tolon, T., Özden, O., Saygi, H., 2015. Farklı sıcaklıklarda deniz hıyarı (*Holothuria tubulosa*, Gmelin, 1788) genç bireylerinin büyüme ve yaşama oranı. Turk. J. Fish. Aquat. Sci. 15, 533–541. https://doi.org/10.4194/1303-2712-v15_2_41.

Günay, D., Tolon, M.T., Emiroğlu, D., 2018. Effects of various stocking densities on hatching and survival rates of sea cucumber *Holothuria tubulosa* eggs (Gmelin, 1788). Ege J. Fish. Aquat. Sci. 35, 381–386. https://doi.org/10.12714/egejfas.2018.35.4.03.

Gvirtzman, G., 1977. Morphology of the Red Sea fringing reefs : a result of the erosional pattern of the last-glacial lowstand sea level and the following holocene recolonization. In: 2ème symp. intern. sur les coraux et récifs coralliens fossi. Proc. 2nd Int. Symp. Corals Foss. Coral Reefs Paris, vol. 8, pp. 480–491.

Hanafy, M.H., 2011. A study on the effect of the sea cucumber *Actinopyga mauritiana* (Echinodermata: Holothuroidea) on the sediment characteristics at El-Gemsha Bay, Red Sea coast, Egypt Mahmoud. Int. J. Environ. Sci. Eng. 2, 35–44.

Hasan, M., 2009. Stock assessment of holothuroid populations in the Red Sea waters of Saudi Arabia. SPC Beche-de-mer Inf. Bull. 29, 31–37.

Hasan, M.H., 2019. Destruction of sea cucumber populations due to overfishing at Abu Ghosoun area, Red Sea. J. Basic Appl. Zool. 80 https://doi.org/10.1186/s41936-019-0074-6.

Hasan, M.H., 2005. Destruction of a *Holothuria scabra* population by overfishing at Abu Rhamada Island in the Red Sea. Mar. Environ. Res. 60, 489–511. https://doi.org/10.1016/j.marenvres.2004.12.007.

Hellal, A.M., 2010. Contribution to the sea cucumber fauna (Echinodermata: Holothuroidea) at the vicinity of Bab El-Mandab, Red Sea, Yemen. Al-Azhar Bull. Sci. 21, 27–68. https://doi.org/10.21608/absb.2010.7352.

Hellal, A.M., Abo Zeid, M.M., El-Sayed, A.A., El-Samra, M., Hassan, M.H., 2007. Zoogeography of the sea cucumber (Holothuroidea: Echinodermata) from the Red Sea. In: The Fourth Scientific Conference on the Environment & Natural Resources. Taiz University, Republic of Yemen.

Henriques, F.F., Serrão, E.A., González-Wangüemert, M., 2016. Novel polymorphic microsatellite loci for a new target species, the sea cucumber *Holothuria mammata*. Biochem. Systemat. Ecol. 66, 109–113. https://doi.org/10.1016/j.bse.2016.03.012.

Ibrahim, D.M., Radwan, R.R., Abdel Fattah, S.M., 2017. Antioxidant and antiapoptotic effects of sea cucumber and valsartan against doxorubicin-induced cardiotoxicity in rats: the role of low dose gamma irradiation. J. Photochem. Photobiol. B Biol. 170, 70–78. https://doi.org/10.1016/j.jphotobiol.2017.03.022.

Ibrahim, H.A.H., 2012. Antibacterial carotenoids of three Holothuria species in Hurghada, Egypt. Egypt. J. Aquat. Res. 38, 185–194. https://doi.org/10.1016/j.ejar.2013.01.004.

Ibrahim, M.Y., Elamin, S.M., Gideiri, Y.B.A., Ali, S.M., 2015. The proximate composition and the nutritional value of some sea cucumber species inhabiting the Sudanese Red Sea. Food Sci. Qual. Manag. 41, 11–17.

Işgören-Emiroğlu, D., Günay, D., 2007. The effect of sea cucumber *Holothuria tubulosa* G. 1788 on nutrient and organic matter contents of bottom sediment of oligotrophy and hypereutrophic shores. Fresenius Environ. Bull. 16, 290–294.

Ismail, H., Lemriss, S., Ben Aoun, Z., Mhadhebi, L., Dellai, A., Kacem, Y., Boiron, P., Bouraoui, A., 2008. Antifungal activity of aqueous and methanolic extracts from the Mediterranean Sea cucumber, *Holothuria polii*. J. Mycol. Med. 18, 23–26. https://doi.org/10.1016/j.mycmed.2008.01.002.

Israel, D., Lupatsch, I., Angel, D.L., 2019. Testing the digestibility of seabream wastes in three candidates for integrated multi-trophic aquaculture: grey mullet, sea urchin and sea cucumber. Aquaculture 510, 364–370. https://doi.org/10.1016/j.aquaculture.2019.06.003.

Kalaeb, T., Ghirmay, D., Semere, Y., Yohannes, F., 2008. Status and preliminary assessment of the sea cucumber fishery in Eritrea. SPC Beche Mer Inf. Bull. 27, 8—12.

Kazzaz, W.M. El, Shabana, E.E., Dar, M.R., Dewedar, A., 2019. The influence of *Holothuria atra* (Echinodermata : Holothuroidea) on bacterial density and sediment characteristics of the Red Sea , Hurghada , Egypt. J. Basic Environ. Sci. 6, 66—77.

Khattab, R.A., Elbandy, M., Lawrence, A., Paget, T., Rae- Rho, J., Binnaser, Y.S., Ali, I., 2018. Extraction, identification and biological activities of saponins in sea cucumber *Pearsonothuria graeffei*. Comb. Chem. High Throughput Screen. 21 https://doi.org/10.2174/1386207321666180212165448.

Khattab, R.M.A., Ali, A.E., El-Nomany, B., Temraz, T.A., 2008. Screening for antibacterial and antifungal activities in some selected marine organisms of the Suez Canal and Red Sea. Egypt. J. Exp. Biol. 4, 223—228.

Künili, İ.E., Çolakoğlu, F.A., 2018. Antioxidant and antimicrobial activity of sea cucumber (*Holothuria tubulosa*, Gmelin 1791) extracts. J. Mar. Sci. Fish. 1 (2), 66—71.

Lawrence, A.J., Afifi, R., Ahmed, M., Khalifa, S., Paget, T., 2010. Bioactivity as an options value of sea cucumbers in the Egyptian Red Sea: contributed paper. Conserv. Biol. 24, 217—225. https://doi.org/10.1111/j.1523-1739.2009.01294.x.

Lawrence, A.J., Ahmed, M., Hanafy, M., Gabr, H., Ibrahim, A., Gab-Alla, A.-F., 2004. Status of the Sea Cucumber Fishery in the Red Sea -the Egyptian Experience. Adv. sea cucumber Aquac. Manag. FAO Fish. Tech. Pap. 463.

Maggi, C., González-Wangüemert, M., 2015. Genetic differentiation among *Parastichopus regalis* populations in the Western Mediterranean Sea: potential effects from its fishery and current connectivity. Mediterr. Mar. Sci. 16, 489—501. https://doi.org/10.12681/mms.1020.

Mahyoub, J.A., Hawas, U.W., Al-Ghamdi, K.M., Aljameeli, M.M.E., Shaher, F.M., Bamakhrama, M.A., Alkenani, N.A., 2016. The biological effects of some marine extracts against *Aedes aegypti* (L.) mosquito vector of the dengue fever in Jeddah Governorate, Saudi Arabia. J. Pure Appl. Microbiol. 10, 1949—1956.

Marquet, N., Conand, C., Power, D.M., Canário, A.V.M., González-Wangüemert, M., 2017. Sea cucumbers, *Holothuria arguinensis* and *H. mammata*, from the southern Iberian Peninsula: variation in reproductive activity between populations from different habitats. Fish. Res. 191, 120—130. https://doi.org/10.1016/j.fishres.2017.03.007.

Massutí, E., Renones, O., 2005. Demersal resource assemblages in the trawl fishing grounds off the Balearic Islands (western Mediterranean). Sci. Mar. 69, 167—181. https://doi.org/10.3989/scimar.2005.69n1167.

Melek, F.R., Tadros, M.M., Yousif, F., Selim, M.A., Hassan, M.H., 2012. Screening of marine extracts for schistosomicidal activity in vitro. Isolation of the triterpene glycosides echinosides A and B with potential activity from the sea Cucumbers *Actinopyga echinites* and *Holothuria polii*. Pharm. Biol. 50, 490—496. https://doi.org/10.3109/13880209.2011.615842.

Mezali, K., Soualili, D.L., 2014. Reproductive cycle of the sea cucumber *Holothuria tubulosa* (Holothuroidea: Echinodermata) in the southwestern Mediterranean Sea. In: Proceedings of the International Congress on Estuaries and Costal Protected Areas"ECPA (İzmir—Turkey, 04—06 November 2014.

Mezali, K., Soualili, D.L., Neghli, L., Conand, C., 2014. Reproductive cycle of the sea cucumber *Holothuria* (platyperona) *sanctori* (Holothuroidea: Echinodermata) in the southwestern Mediterranean Sea: interpopulation variability. Invertebr. Reprod. Dev. 58, 179—189. https://doi.org/10.1080/07924259.2014.883337.

Mezali, K., Thandar, A., Zootaxa, I.K., 2020. Paraleptopentacta, a new Mediterranean and north-west Atlantic Sea cucumber genus, with the first record of *P. tergestina* n. comb.(Echinodermata: Dendrochirotida. Zootaxa 2, 199—210.

Mezali, K., Thandar, A.S., 2014. First record of *Holothuria* (Roweothuria) *arguinensis* (Echinodermata: Holothuroidea: Aspidochirotida: Holothuriidae) from the Algerian coastal waters. Mar. Biodiversity Rec. 7 https://doi.org/10.1017/S1755267214000438.

Mohamed, A.S., Mahmoud, S.A., Soliman, A.M., Fahmy, S.R., 2019. Antitumor activity of saponin isolated from the sea cucumber, *Holothuria arenicola* against ehrlich ascites carcinoma cells in swiss albino mice. Nat. Prod. Res. 1—5. https://doi.org/10.1080/14786419.2019.1644633.

Mona, M.H., Omran, N.E.E., Mansoor, M.A., El-Fakharany, Z.M., 2012. Antischistosomal effect of holothurin extracted from some Egyptian sea cucumbers. Pharm. Biol. 50, 1144—1150. https://doi.org/10.3109/13880209.2012.661741.

Moussa, R.M., Wirawati, I., 2018. Observations on some biological characteristics of *Holothuria polii* and *Holothuria sanctori* from Mediterranean Egypt. Int. J. Fish. Aquat. Stud. 6, 351—357.

Neofitou, N., Lolas, A., Ballios, I., Skordas, K., Tziantziou, L., Vafidis, D., 2019. Contribution of sea cucumber *Holothuria tubulosa* on organic load reduction from fish farming operation. Aquaculture 501, 97—103. https://doi.org/10.1016/j.aquaculture.2018.10.071.

Ocaña, A., De la Morena, I., Moriana, M., Alonso, M.R., Ibáñez, M., 1982. Algunos Equinodermos de la Costa de Málaga (Mar de Alborán). Investig. Pesq. 46, 433—442.

Ocaña, A., Sánchez Tocino, L., 2005. Spawning of *Holothuria tubulosa* (Holothurioidea , Echinodermata) in the Alboran Sea (Mediterranean Sea). Zool. Baetica 16, 147—150.

Omar, H.A., Abdel Razek, F.A., Abdel Rahman, S.H., El Shimy, N.A., 2013. Reproductive periodicity of sea cucumber *Bohadschia vitiensis* (Echinodermata: Holothuroidea) in Hurghada area, Red Sea, Egypt. Egypt. J. Aquat. Res. 39, 115—123. https://doi.org/10.1016/j.ejar.2013.06.002.

Omran, N., 2006. Fungicidal and bactericidal effects of holothurin extracted from four species of sea cucumber inhabiting Mediterranean Sea and Red Sea coasts of Egypt. Proceeding 4th Int. Conf. Biol. Sci. Zool. 4, 1—6.

Omran, N.E., Allam, N.G., 2013. Screening of microbial contamination and antimicrobial activity of sea cucumber *Holothuria polii*. Toxicol. Ind. Health 29, 944—954. https://doi.org/10.1177/0748233712448116.

Omran, N.E., Khedr, A.M., 2015. Structure elucidation, protein profile and the antitumor effect of the biological active substance extracted from sea cucumber *Holothuria polii*. Toxicol. Ind. Health 31, 1—8. https://doi.org/10.1177/0748233712466135.

Omran, N.E.S.E.S., 2013. Nutritional value of some Egyptian sea cucumbers. Afr. J. Biotechnol. 12, 5466—5472. https://doi.org/10.5897/ajb2013.13020.

Özer, N.P., Mol, S., Varlık, C., 2005. Effect of the handling procedures on the chemical composition of sea cucumber. Turk. J. Fish. Aquat. Sci. 74, 71—74.

Öztürk, F., Gündüz, H., 2018. The effect of different drying methods on chemical composition, fatty acid, and amino acid profiles of sea cucumber (*Holothuria tubulosa* Gmelin, 1791). J. Food Process. Preserv. 42 https://doi.org/10.1111/jfpp.13723.

Piccinno, M., Schiavone, R., Zilli, L., Sicuro, B., Storelli, C., Vilella, S., 2013. Sea cucumber meal as alternative protein source to fishmeal in gilthead sea bream (*Sparus aurata*) nutrition: effects on growth and welfare. Turk. J. Fish. Aquat. Sci. 13, 305—313. https://doi.org/10.4194/1303-2712-v13_2_12.

Pinardi, N., Arneri, E., Crise, A., Ravaioli, M., Zavatarelli, M., 2006. The physical, sedimentary and ecological structure and variability of shelf areas in the Mediterranean Sea (27). Sea 14, 1243–1330.

Pomin, V.H., 2014. Holothurian fucosylated chondroitin sulfate. Mar. Drugs 12 (1), 232–254. https://doi.org/10.3390/md12010232.

Rakaj, A., Fianchini, A., Boncagni, P., Lovatelli, A., Scardi, M., Cataudella, S., 2018. Spawning and rearing of *Holothuria tubulosa*: a new candidate for aquaculture in the Mediterranean region. Aquacult. Res. 49, 557–568. https://doi.org/10.1111/are.13487.

Rakaj, A., Fianchini, A., Boncagni, P., Scardi, M., Cataudella, S., 2019. Artificial reproduction of *Holothuria polii*: a new candidate for aquaculture. Aquaculture 498, 444–453. https://doi.org/10.1016/j.aquaculture.2018.08.060.

Ramón, M., Lleonart, J., Massutí, E., 2010. Royal cucumber (*Stichopus regalis*) in the northwestern Mediterranean: distribution pattern and fishery. Fish. Res. 105, 21–27. https://doi.org/10.1016/j.fishres.2010.02.006.

Razek, F.A.A., Mona, M.H., Rahman, S.H.A., El, M.M., Moussa, R.M., Taha, S.M., 2010. Observations on the abundance of holothurian species along the Alexandria coast, Egypt. In: Proc. 39th CIESM Congr. May 10–14, 2010, Italy 2006, pp. 1–9.

Razek, F.A.A., Rahman, S.H.A., Moussa, R.M., Mena, M.H., El-Gamal, M.M., 2012. Captive spawning of *Holothuria arenicola* (Semper, 1868) from Egyptian Mediterranean coast. Asian J. Bio. Sci. 5, 425–431. https://doi.org/10.3923/ajbs.2012.425.431.

Roggatz, C.C., González-Wangüemert, M., Pereira, H., Vizetto-Duarte, C., Rodrigues, M.J., Barreira, L., da Silva, M.M., Varela, J., Custódio, L., 2018. A first glance into the nutritional properties of the sea cucumber *Parastichopus regalis* from the Mediterranean Sea (SE Spain). Nat. Prod. Res. 32, 116–120. https://doi.org/10.1080/14786419.2017.1331224.

Schillaci, D., Cusimano, M.G., Cunsolo, V., Saletti, R., Russo, D., Vazzana, M., Vitale, M., Arizza, V., 2013. Immune mediators of sea-cucumber *Holothuria tubulosa* (Echinodermata) as source of novel antimicrobial and anti-staphylococcal biofilm agents. AMB Express 3, 1–10. https://doi.org/10.1186/2191-0855-3-35.

Sellem, F., Brahmi, Z., Mnasser, H., Rafrafi, S., Bouhaouala-Zahar, B., 2017. Antimicrobial activities of coelomic fluid and body wall extracts of the edible Mediterranean sea cucumber *Holothuria tubulosa* Gmelin, 1790. Cah. Biol. Mar. 58, 181–188. https://doi.org/10.21411/CBM.A.479839A5.

Sezgin, M., Şahin, F., Bat, L., 2007. Presence of Stereoderma kirschbergi (Echinodermata: Holothuroidea) on Sinop Peninsula coast, Turkey: first record from Turkish Black Sea. JMBA2-Biodiversity Records 10.

Shabana, E.E., El Kazzaz, W., Dar, M., Dewedar, A., 2018. The influence of *Holothuria atra* (Echinodermata: Holothuroidea) on organic matter assimilation, Ammonification and nitrification rate of sediment. Hurghada, Red Sea, Egypt. J. Sci. Res. Sci. 35, 1–15. https://doi.org/10.21608/jsrs.2018.25515.

Sicuro, B., Piccinno, M., Gai, F., Abete, M.C., Danieli, A., Dapra, F., Mioletti, S., Vilella, S., 2012. Food quality and safety of Mediterranean Sea cucumbers *Holothuria tubulosa* and *Holothuria polii* in southern Adriatic Sea. Asian J. Anim. Vet. Adv. 7, 851–859. https://doi.org/10.3923/ajava.2012.851.859.

Silchenko, A.S., Stonik, V.A., Avilov, S.A., Kalinin, V.I., Kalinovsky, A.I., Zakharenko, A.M., Smirnov, A.V., Mollo, E., Cimino, G., 2005. Holothurins B2, B3, and B4, new triterpene glycosides from Mediterranean Sea cucumbers of the genus Holothuria. J. Nat. Prod. 68, 564–567. https://doi.org/10.1021/np049631n.

Slimane-Tamacha, F., Soualili, D., Mezali, K., 2019. Reproductive biology of *Holothuria* (Roweothuria) *poli* (Holothuroidea: Echinodermata) from Oran Bay, Algeria. SPC Beche-de-mer Inf. Bull. 39, 47—53.

Soliman, Y.A., Ibrahim, A.M., Tadros, H.R.Z., Abou-Taleb, E.A., Moustafa, A.H., Hamed, M.A., 2016. Antifouling and antibacterial activities of marine bioactive compounds extracted from some Red Sea sea cucumber. Int. J. Contemp. Appl. Sci. 3, 83—103.

Tolon, T., Emiroğlu, D., Günay, D., Hancı, B., 2017a. Effect of stocking density on growth performance of juvenile sea cucumber *Holothuria tubulosa* (Gmelin, 1788). Aquacult. Res. 48, 4124—4131. https://doi.org/10.1111/are.13232.

Tolon, M.T., Emiroglu, D., Gunay, D., Ozgul, A., 2017b. Sea cucumber (*Holothuria tubulosa* Gmelin, 1790) culture under marine fish net cages for potential use in integrated multitrophic aquaculture (IMTA). Indian J. Geo-Mar. Sci. 46, 749—756.

Tolon, M.T., Engin, S., 2019. Gonadal development of the holothurian *Holothuria polii* (Delle Chiaje, 1823) in spawning period at the Aegean Sea (Mediterranean Sea). Ege J. Fish. Aquat. Sci. 36, 379—385. https://doi.org/10.12714/egejfas.36.4.09.

Tolon, T., Emiroğlu, D., Günay, D., Saygı, H., 2015. Sediment tane boyutunun deniz hıyarı (*Holothuria tubulosa*) genç bireylerinin büyüme performansı üzerine etkileri. Turk. J. Fish. Aquat. Sci. 15, 555—559. https://doi.org/10.4194/1303-2712-v15_2_43.

Toscano, A., Cirino, P., 2018. First evidence of artificial fission in two Mediterranean species of holothurians: *Holothuria tubulosa* and *holothuria polii*. Turk. J. Fish. Aquat. Sci. 18, 1141—1145. https://doi.org/10.4194/1303-2712-v18_10_01.

Turk Culha, S., Dereli, H., Karaduman, F.R., Culha, M., 2016. Assessment of trace metal contamination in the sea cucumber (*Holothuria tubulosa*) and sediments from the Dardanelles Strait (Turkey). Environ. Sci. Pollut. Res. 23, 11584—11597. https://doi.org/10.1007/s11356-016-6152-0.

Vafeiadou, A.M., Antoniadou, C., Vafidis, D., Fryganiotis, K., Chintiroglou, C., Density, K., Sea, A., 2010. Density and biometry of the exploited holothurian *Holothuria tubulosa* at the Dodecanese, South Aegean Sea. Rapp. Comm. Int. Mer Médit 39, 691.

Vaipulu, S.'E.U., 2009. Design a Prototype Solar Dryer for Drying Sea Cucumber. University of Southern Queensland.

Valente, S., Serrão, E.A., González-Wangüemert, M., 2015. West versus East Mediterranean Sea: origin and genetic differentiation of the sea cucumber *Holothuria polii*. Mar. Ecol. 36, 485—495. https://doi.org/10.1111/maec.12156.

Valls, A., 2004. Observation of natural spawning of *Holothuria tubulosa*. Beche-de-Mer Inf. Bull. 19, 40.

Van Dyck, S., Gerbaux, P., Flammang, P., 2009. Elucidation of molecular diversity and body distribution of saponins in the sea cucumber *Holothuria forskali* (Echinodermata) by mass spectrometry. Comp. Biochem. Physiol. B Biochem. Mol. Biol. 152, 124—134. https://doi.org/10.1016/j.cbpb.2008.10.011.

Yucel-Gier, G., Kucuksezgin, F., Kocak, F., 2007. Effects of fish farming on nutrients and benthic community structure in the Eastern Aegean (Turkey). Aquacult. Res. 38, 256—267. https://doi.org/10.1111/j.1365-2109.2007.01661.x.

Sea cucumbers research in the Persian Gulf

5.1 Region under study

The Persian Gulf, or the Arabian Gulf, is a shallow prolongation of the Indian Ocean, which gives access to Europe through the Gulf of Oman to which it is joined by the Strait of Hormuz (Fig. 5.1). It is an almost inland sea with a maximum length of about 990 km and a maximum depth of 90 m (average of 36 m). It occupies a surface

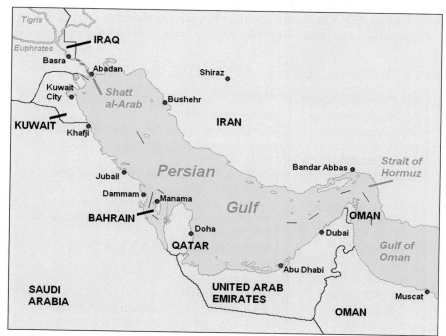

FIGURE 5.1

Map of the Persian Gulf.

Credit: Hégésippe Cormier aka Hégésippe CC BY-SA 3.0 via Wikimedia Commons.

area of about 239,000 km^2. Eight countries have coasts along the Persian Gulf: the United Arab Emirates, Oman, Saudi Arabia, Qatar, Bahrain, Kuwait and Iraq on one side and Iran on the other side (Bower et al., 2000; Kämpf and Sadrinasab, 2006).

5.2 Biology and ecology of sea cucumber species

Despite the growing number of studies focussing on holothurians worldwide, the Persian sea cucumbers have not received much attention. Heding (1940) recorded 17 species of holothurians in the waters around Iran (Dabbagh et al., 2011; Heding, 1940). Recently, researchers reported 15 sea cucumber species found in the Persian Gulf (Table 5.1) and mostly collected from the shallow water. Among them, the sea cucumber *Holothuria scabra* is the most exploited sea cucumber species (Dabbagh et al., 2012a).

5.2.1 *Stichopus hermanni* (Brandt, 1835)

The curry sea cucumber *Stichopus hermanni* was found on coral reefs between 4 and 8 m deep. This species has a brown-yellow colour and is mottled with papillae that darken in colour (Fig. 5.2). Its dorsal surface has papillae placed on small warts, and the ventral surface has numerous podia, roughly cylindrical and higher in number

Table 5.1 Holothurians identified recently from the Persian Gulf.

	Sea cucumber species	References
1	*Ohshimella ehrenbergii*	Dabbagh et al. (2012b).
2	*Stichopus hermanni*	Salarzadeh et al. (2013), Tehranifard et al. (2006)
3	*Stichopus monotuberculatus*	Maryam et al. (2012), Salarzadeh et al. (2013)
4	*Holothuria arenicola*	Mehrdost et al. (2015), Salari-Aliabadi and Monjezi-Veysi (2020)
5	*Holothuria leucospilota*	Ghobadyan et al. (2012), Salarzadeh et al. (2013)
6	*Holothuria parva*	Pourvali et al. (2014), Salari-Aliabadi and Monjezi-Veysi (2020)
7	*Holothuria bacilla*	Afkhami et al. (2015), Mehrdost et al. (2015)
8	*Holothuria insignis*	Afkhami et al. (2015)
9	*Holothuria* sp.	Pourvali et al. (2014)
10	*Holothuria pardalis*	Pourvali et al. (2014)
11	*Holothuria scabra*	Pourvali et al. (2014), Salarzadeh et al. (2013))
12	*Holothuria hilla*	Afkhami et al. (2012a), Salarzadeh et al. (2013)
13	*Holothuria impatiens*	Afkhami et al. (2012b), Salarzadeh et al. (2013)
14	*Holothuria notabilis*	Afkhami et al. (2014)
15	*Holothuria cinerascens*	(Fatemi et al. (2011)

FIGURE 5.2

The sea cucumber *Stichopus hermanni.*

Credit: François Michonneau CC BY 3.0 via Wikimedia Commons.

than the papillae, and it does not have Cuvierian tubules. The ossicles were found in the form of tables, rosettes and C-shaped rods. Twenty leaf-shaped tentacles surround the subterminal mouth (Tehranifard, 2011).

Tehranifard et al. (2006) conducted a study on the reproductive biology of the sea cucumber *S. hermanni* from Kish Island in the Persian Gulf. The colour of the gonads is similar to other sea cucumber populations, and the spawning was found in coincidence with high water temperature (Table 5.2). Also, Ghobadyan et al. (2012) showed that sea cucumbers sampled monthly, except December and January, along the Bustaneh Coast of the Persian Gulf have five maturity stages. The tubules were found to be much longer and narrower in males than in females. In a study conducted to examine the feeding regime of the sea cucumber *S. hermanni* in Qeshm Island, Rashidi et al. (2017) showed that the diet of the sea cucumber *S. hermanni* in the southeast of the Qeshm Island consists mainly of diatoms, blue-green algae and microbenthic animals (i.e. nematodes, foraminifers and gastropod) (Rashidi et al., 2017).

Table 5.2 Reproductive characters of the sea cucumber *Stichopus hermanni* collected from the Persian Gulf.

Sea cucumber species	Spawning period	Mature oocyte diameter	Size at first maturity	Relative fecundity	Absolute fecundity	Reference
Stichopus hermanni	July —August	200 μm	310 mm (average length)	8 × 103 oocytes.	6—10 × 103	Tehranifard et al. (2006)

Credit: Tehranifard, A., Uryan, S., Vosoghi, G., 2006. Reproductive cycle of Stichopus herrmanni from Kish Island, Iran. Beche Mer Bull. 24, 22—27.

5.2.2 *Holothuria leucospilota* (Brandt, 1835)

The adults of sea cucumber *Holothuria leucospilota* were collected by snorkelling in depths of 0.5—1.5 m at low tide (Dabbagh et al., 2011). The body of this species is elongated, smooth and relatively thin, bottle-shaped and black in colour (Fig. 5.3). Twenty tentacles surround the ventral mouth, and the anus is terminal. The Cuvierian tubules exist and can be quickly ejected when the animal is disturbed. The ossicles were found in the form of rods in the tentacles, in the form of tables and buttons in the body wall, and long rods were found in the dorsal podia (Majid et al., 2012; Pourvali et al., 2014; Salarzadeh et al., 2013).

From the Persian Gulf, the sea cucumbers *H. leucospilota* was observed to have five maturity stages (Ghobadyan et al., 2012): early growth, growth, advanced growth, mature and post-spawning. The tubules in the males were found to be much longer and narrower than in females. The sex cannot be distinguished externally but after dissection, as the gonads differ in colour between males and females (Ghobadyan et al., 2012).

5.2.3 *Holothuria scabra* (Jaeger, 1833)

The sea cucumber *H. scabra* was collected from muddy beds at a depth of 5 m. Their body is grey-black ventrally with white or yellow horizontal bands, and tiny dark dots were found on the ventral surface (Fig. 5.4). Twenty short tentacles surround the ventral mouth, and Cuvierian tubules exist. The ossicles were found in the form of rods in the tentacles. Tables and buttons ossicles were found in the dorsal body wall. The ventral body wall has branched rods and buttons. The papillae have tables, buttons and perforated rods (Dabbagh and Sedaghat, 2012; Pourvali et al., 2014).

FIGURE 5.3

The sea cucumber *Holothuria leucospilota*.

Credit: By Philippe Bourjon Licensed under CC BY-SA 3.0 via Wikimedia Commons.

FIGURE 5.4

The sea cucumber *Holothuria scabra*.

Credit: Ahmed (2009).

5.2.4 *Holothuria hilla* (Lesson, 1830)

The sea cucumber *Holothuria hilla* was collected at depths of 5–15 m (Salarzadeh et al., 2013) and 0–20 m (Afkhami et al., 2012a). The body is reddish-brown, characterised by the presence of numerous white spots marking the base of creamy white large conical papillae (Fig. 5.5). Twenty small yellowish tentacles surround the ventral mouth. The ossicles in the body wall were found in the form of tables and buttons (Afkhami et al., 2012a; Salarzadeh et al., 2013).

FIGURE 5.5

The sea cucumber *Holothuria hilla*.

Credit: Philippe Bourjon. Licensed under CC BY SA 3.0 via Wikimedia Commons.

5.2.5 *Holothuria impatiens* (Forsskål, 1775)

The sea cucumber *Holothuria impatiens* was observed among rocks in shallow water at a depth of 0–2 m, but they also can be observed up to 30 m (Afkhami et al., 2012b). The body of this species is bottle-shaped, rough and light brown in colour. The dorsal surface has conical warts from which filamentous appendages appear. Four to five dark brown transverse bands appear on the upper side near the anterior end (Fig. 5.6). The ventral surface is beige with brown dorsal spots more or less dark, and the tube feet have bright areas. Twenty tentacles surround the ventral mouth. The anus is terminal, and the Cuvierian tubules exist. The ossicles were found in the shape of buttons and tables (Afkhami et al., 2012b).

5.2.6 Other species

The sea cucumber *Stichopus monotuberculatus* was reported from the Persian Gulf. The dorsal side has an orange-brown to grey-green colour with dark green to black spots. The ventral side has a grey-green colour with small dark spots. Twenty large tentacles surround the mouth. The tube feet are yellowish-brown in ambulacral areas, and large conical papillae are distributed more or less randomly on the dorsal surface; 8–10 larger papillae are found laterally (Maryam et al., 2012).

The sea cucumber *Ohshimella ehrenbergii* was collected at depths up to 2.5 m between the cavity of rocks (Dabbagh et al., 2012b). The body of this species was found to be greyish-brown, and the ventral side is lighter than the dorsal side. The Cuvierian organs were not found, and 20 whitish tentacles surround the ventral mouth. The ossicles were found in the form of rods in the dorsal body wall and the ventral body wall in the form of rods, plates and rosettes. In the tube feet, the ossicles were found in the form of rods, plates and rosettes. In the tentacles, the ossicles were found in the form of rods (Dabbagh et al., 2012b).

FIGURE 5.6

The sea cucumber *Holothuria impatiens*.

Credit: Philippe Bourjon. Licensed under CC BY 4.0 via Wikimedia commons.

The sea cucumber *Holothuria bacilla* was collected from shallow water between cavities in the Persian Gulf. The body of this species is brown, and 20 short tentacles surround the ventral mouth. The ventral body has large podia and is stiff throughout the trivium, whereas the dorsal podia are thinner and scattered. The ossicles were found in the form of rods and plates (Afkhami et al., 2015).

The sea cucumber *H. insignis* was found in the shallow water under rocks or the sand in the Persian Gulf. The body is brown or dark grey with papillae scattered and whitish spots distributed throughout the body. Around 17—20 tentacles surround the ventral mouth. The anus is terminal and surrounded by five papillae. Cuvierian tubules exist. The ossicles were found in the form of rods, buttons and plates (Afkhami et al., 2015).

The sea cucumber *Holothuria arenicola* was collected from the Persian Gulf. The body is roughly cylindrical and creamy white and sometimes yellow or red with two rows of black spots on the dorsal surface. The mouth is central, and the anus is subterminal or terminal. The ossicles were observed in the tentacles in the form of tables, rods and buttons. In the anterior part, the ossicles were observed in the form of buttons and tables. In the dorsal part, the ossicles were observed in the form of buttons. In the anal part, the ossicles were observed in the form of buttons, tables and rods. In the ventral part, the ossicles were observed in the form of buttons and tables (Salari-Aliabadi and Monjezi-Veysi, 2020). Pourvali et al. (2014) found that the sea cucumbers *H. arenicola* and *H. leucospilota* have similar table spicule shape.

The sea cucumber *Holothuria parva* was collected from the Persian Gulf. This species has a cylindrical body brown in colour, with yellow tube feet. It was also reported as dark green to black in colour. The number of the papillae was found smaller than the number of the tube feet on the ventral surface. Twenty cryptic tentacles surround the ventral mouth. Ossicles were found in the form of rods in the dorsal, ventral body wall and the tentacles. Plates were also found in the ventral part (Ehsanpour et al., 2016; Salari-Aliabadi and Monjezi-Veysi, 2020).

Research on population genetics in the Persian Gulf showed that the population of the sea cucumber *H. parva* between two regions in the Persian Gulf (Dayer and Bostaneh) has high gene flow and little genetic differentiation, which suggests common ancestors and the stock of the sea cucumber can be replenished from many resources (Ali-abadi, 2016). Genetic characterisation analysis conducted on the sea cucumber *H. parva* through sequencing segments of the mitochondrial 16S RNA and cytochrome oxidase (I) confirmed its identity with previously identified one (Ehsanpour et al., 2016).

The sea cucumber *Holothuria notabilis* was collected from sandy habitat between seagrass beds at 8—15 m depth in the Persian Gulf. The body of this species is cylindrical with dark brown or black dots arranged on the dorsal surface in two rows of eight dots, and 20 yellow tentacles surround the ventral mouth. The Cuvierian tubules inside the body are white and rarely expelled. The ossicles were found in the form of numerous small nodulous buttons and a few tables. In the dorsal body wall, tables with larger disc diameter and with fully developed spire were found (Afkhami et al., 2014) (Table 5.3).

Table 5.3 Habitat preference for Persian sea cucumber.

Species	Habitat	References
Ohshimella ehrenbergii	Collected at depths up to 2.5 m from rocky shore habitats between the cavity of rocks.	Dabbagh et al. (2012b).
Stichopus hermanni	Inhabits shallow water and common in reef habitats. It is widely distributed across the tropical Indo-Pacific region.	Tehranifard (2011).
Holothuria arenicola	It is found at depths of 0–7 m and depth of 30 m on sandy bottoms, rocks, coral reefs, sandy beaches and rock-sandy beaches in subtidal areas.	Mehrdost et al. (2015), Pourvali et al. (2014)
Holothuria leucospilota	Found on sediment and so was appeared on substrata. Muddy and rocky substrate and was collected by snorkelling in depths of 0.5–1.5 m at low tide.	Dabbagh et al. (2011), Pourvali et al. (2014)
Holothuria parva	Common around Hormuz Island. Found usually in substrata, which included mud, mainly under rock, occasionally in deep bed (5–10 cm).	Pourvali et al. (2014)
Holothuria bacilla	Common around Hormuz Island; found usually in substrata, which included mud, mainly under rock, occasionally in deep bed (5–10 cm). inhabits shallow waters and lives under rock fragments.	Mehrdost et al. (2015), Pourvali et al. (2014)
Holothuria sp.	Was found only in winter.	Pourvali et al. (2014)
Holothuria pardalis	Found in muddy and rocky habitats.	Pourvali et al. (2014)
Holothuria scabra	Appears on substrata collected from sandy and muddy habitats.	Pourvali et al. (2014)

5.3 Aquaculture development

The sea cucumber *H. leucospilota* collected from the coastal western Persian Gulf was studied for its development and growth. The broodstock was stored in 38–40 ppt and 33°C, and the spawning was induced by thermal stimulation. The embryos develop to auricularia larvae after 120 h from fertilisation (Soltani et al., 2010). Likewise, *H. leucospilota* from the northern Persian Gulf collected during the summer induced to spawn by combining water pressure and thermal stimulation (Chapter 6). Larvae were fed using unicellular algae and *Sargassum* extract. Early juveniles were obtained on day 33. The survival rate was 4.2% for the juvenile stage. With larvae reared at 40 ppt, doliolaria larvae were only obtained by day 22. Preparing the settlement plates with *Sargassum* extract and determining the density of the *Sargassum* extract were problems encountered in the project (Dabbagh et al., 2011).

Furthermore, Dabbagh and Sedaghat (2012) conducted a trial to breed the sea cucumber *H. scabra*, which was collected from Qeshm Island. The sea cucumber *H. scabra* was collected in June and stored in a water temperature of 24°C. The broodstock was induced for spawning in September by combining the water pressure and temperature simulation. Early juveniles were first observed on day 24. Juveniles that reared in a large tank with natural sunlight reached larger sizes than juveniles under artificial light. Larvae were kept at 26°C and fed with unicellular algae. Juveniles fed on extracts of *Sargassum* and *Padina* reached larger sizes than those fed with other foods. Juveniles had better growth under the partial shade with natural sunlight than indoors. However, after 1 year, juveniles averaged only 22 g in weight. The high water temperature of over 30°C year-round and high density were the major problems encountered in the production of the sea cucumber in this area (Dabbagh and Sedaghat, 2012).

5.4 Sea cucumbers utilisation
5.4.1 Nutritional and medicinal values
5.4.1.1 Nutritional values
The Persian sea cucumbers have high nutritional values. Yahyavi et al. (2012) reported that the sea cucumbers *H. scabra* and *H. leucospilota* from Qeshm Island (Persian Gulf) are rich in palmitic acid (Fig. 5.7) and arachidonic acid (Fig. 5.8)

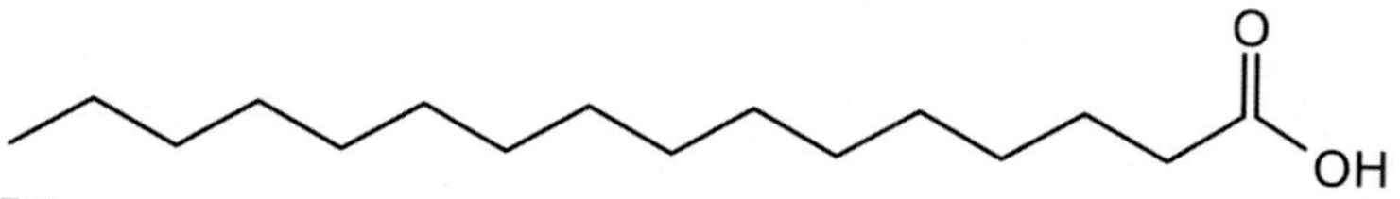

FIGURE 5.7

Chemical structure of palmitic acid.

Credit: BartVL71. Licensed under CC BY-SA 3.0 via Wikimedia commons.

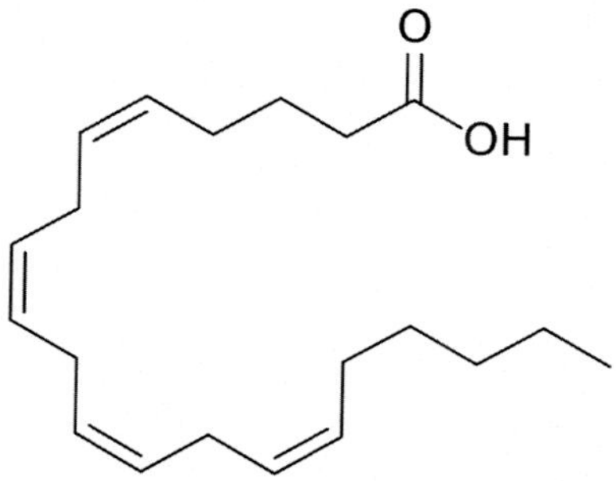

FIGURE 5.8

Chemical structure of arachidonic acid.

Credit: Edward the Confessor. Licensed under CC BY-SA 4.0 via Wikimedia commons.

of saturated fatty acid (SFA) and polyunsaturated fatty acids. The primary monoun-saturated fatty acids in *H. scabra* and *H. leucospilota* were gadoleic acid (Fig. 5.9) and cis-oleic acid (Fig. 5.10), respectively (Yahyavi et al., 2012). Among 19 identi-fied fatty acids obtained from *H. scabra* oil, heneicosanoic acid (Fig. 5.11), linoleic acid (Fig. 5.12), palmitic acid (Fig. 5.7), stearic acid (Fig. 5.13) and myristic acid (Fig. 5.14) have the highest values. The total amount of unsaturated fatty acid and SFA in the sea cucumber *H. scabra* is 59.194% and 40.806%, respectively (Jadavi et al., 2015). Furthermore, the sea cucumber *H. leucospilota* is rich in bioactive com-pounds such as fatty acids from the methanolic extract. Moreover, the nutritional composition of sea cucumbers varies between species. Salarzadeh et al. (2012) re-ported the nutritional composition of the sea cucumbers *H. arenicola* and

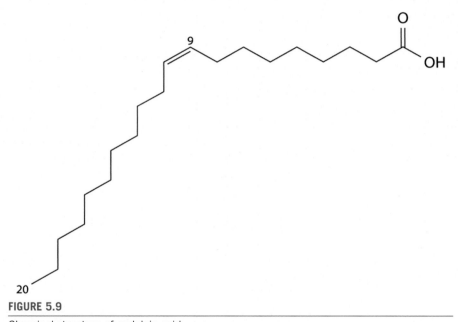

FIGURE 5.9

Chemical structure of gadoleic acid.

Credit: Public domain via Wikimedia commons.

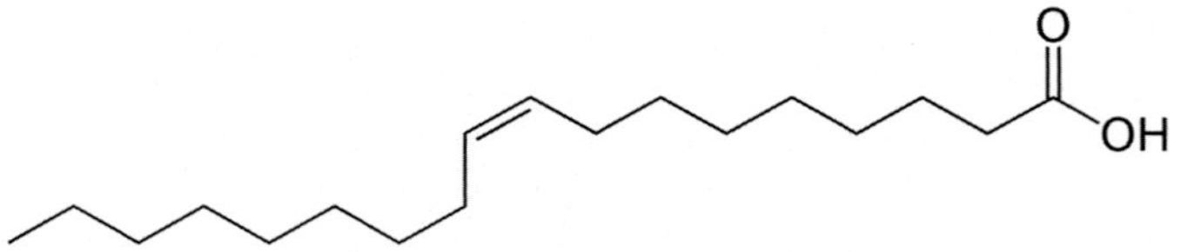

FIGURE 5.10

Chemical structure of oleic acid.

Credit: Public domain via Wikimedia commons.

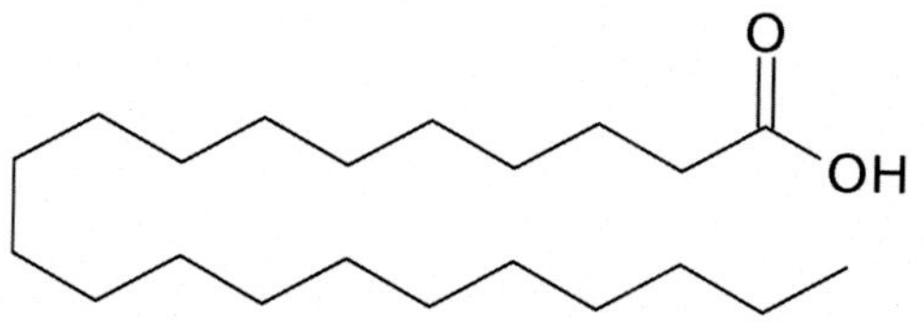

FIGURE 5.11

Chemical structure of heneicosanoic acid.

Credit: Public domain via Wikimedia commons.

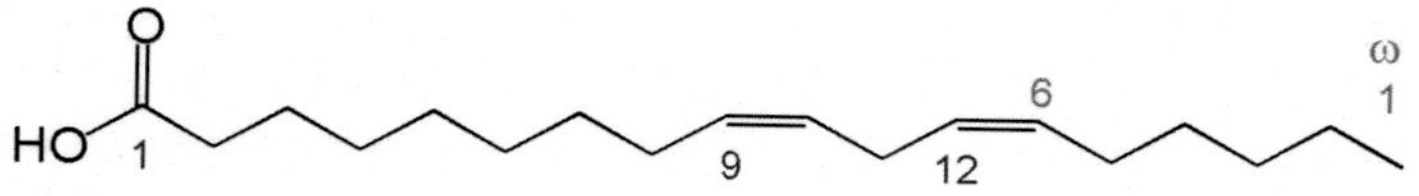

FIGURE 5.12

Chemical structure of linoleic acid.

Credit: Public domain via Wikimedia commons.

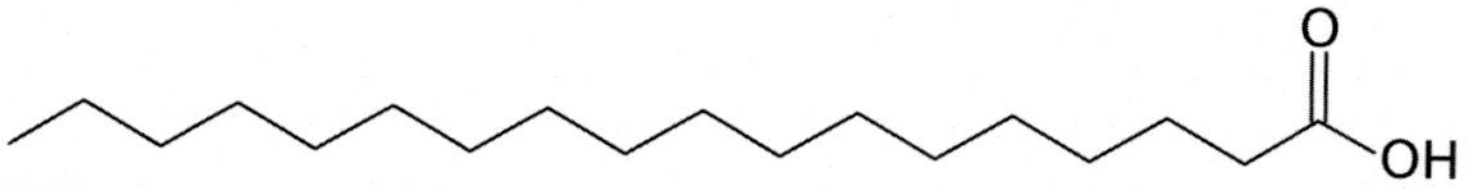

FIGURE 5.13

Chemical structure of stearic acid.

Credit: Public domain via Wikimedia commons.

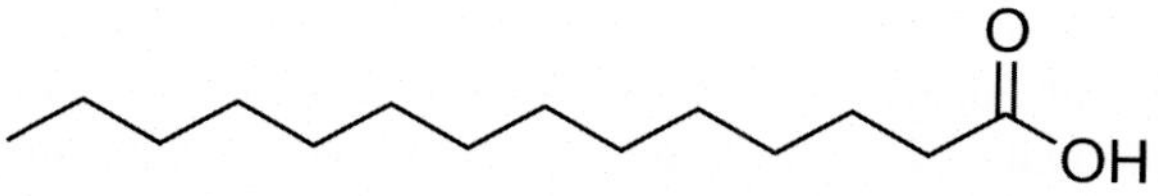

FIGURE 5.14

Chemical structure of myristic acid.

Credit: Public domain via Wikimedia commons.

H. parva collected from the Persian Gulf. Ash content in *H. parva* was higher than *H. arenicola*; however, protein and crude fibre contents in *H. arenicola* were higher than *H. parva*.

5.4.1.2 Anti-microbial properties

Extracts from the Persian sea cucumber have been proven as potential anti-bacterial and anti-fungal agents. The body wall extract of the sea cucumber *H. leucospilota* showed anti-bacterial activity. The aqueous, phosphate-buffered saline extract did not show any anti-bacterial or inhibitory activity. However, the chloroform extract demonstrated high levels of anti-bacterial activity against *Streptococcus salivarius* and exhibited low levels of activity against *Streptococcus mutans*. Also, the hexane and methanolic extracts were found with no anti-bacterial activity against *S. mutans* but exhibited anti-bacterial activity against *S. salivarius* (Kiani et al., 2014). In another study, the methanol, chloroform and hexane extracts of the body wall, gonads and intestine from the sea cucumber *H. leucospilota* were investigated against the bacteria *Escherichia coli* (Farjami et al., 2014b). Chloroform extracts showed anti-bacterial activity at concentrations of 5 and 10 mg per ml; however, methanol extracts had no effect. The hexane extract of body wall at concentrations of 5 mg and 10 mg per ml and the hexane extract of intestines at a concentration of 2.5, 5 and 10 mg per ml have anti-bacterial activity. None of the concentrations of the gonadal hexane extract showed any anti-bacterial activity. Only the hexane extract of the intestines killed the bacteria at a concentration of 10 mg/mL (Farjami et al., 2014b). In another anti-microbial activity research on the sea cucumber *H. leucospilota*, Ghadiri et al. (2018) used phosphate-buffered saline, ethanol and acetonitrile solvents to evaluate extract of the body wall, respiratory tree and gastrointestinal tract of the sea cucumber against the bacteria *S. mutans* and *Streptococcus sobrinus* and the fungi *Candida parapsilosis*, *Candida albicans* and *Candida glabrata* (Ghadiri et al., 2018). The researchers found that the extracts of the respiratory tree with acetonitrile and phosphate-buffered saline solvents showed the highest effect against *C. albicans*, *C. parapsilosis* and *C. glabrata*. Except for the ethanol extract of the respiratory tree, all extracts of the respiratory tree can inhibit the growth of *S. sobrinus*. Also, the extract of the respiratory tree had the maximum anti-fungal effect against *C. albicans* in comparison with the other tissues (Ghadiri et al., 2018). Furthermore, the water-methanol extracts from the body wall of *H. leucospilota* collected from the Persian Gulf possess anti-bacterial and anti-fungal activity (Adibpour et al., 2014).

From the sea cucumber *H. parva*, Ebrahimi et al. (2018) demonstrated that the crude methanol and ethanol extracts of the sea cucumber *H. parva* have anti-microbial activity against *E. coli*, *Pseudomonas aeruginosa* and *Enterococcus faecalis*. The Fourier-transform infrared spectroscopy spectra showed the presence of five components: ouabain, glycerol, gluconic acid, spectinomycin and capreomycin (Ebrahimi et al., 2018). Likewise, ethanolic and methanolic extracts of *H. parva* indicated anti-bacterial activity against *E. coli*, *Vibrio parahaemolyticus* and *Staphylococcus aureus* at concentrations higher than 20 µg/mL, and minimum bactericidal concentration was observed for ethanolic and methanolic extracts at 500 mg/µL against *E. coli* (Shadi and Oujifard, 2019).

Anti-microbial activity of the organic solvents from different parts of the sea cucumbers *H. parva*, *H. scabra* and *H. leucospilota* were evaluated (Mashjoor and Yousefzadi, 2017). The researchers showed that except for *C. albicans* (ATCC 10231), *P. aeruginosa* (ATCC 85327) and *Klebsiella pneumonia* (ATCC 10031), all the sea cucumbers extracts confirmed their strong potential antagonistic effect against the bacterial and fungal indicators, *Bacillus subtilis* (ATCC 465), *Bacillus pumilus* (PTCC 1274), *E. faecalis* (ATCC 29737), *S. aureus* (ATCC 25923), *Staphylococcus epidermidis* (ATCC 12228), *E. coli* (ATCC 25922), *K. pneumoniae* (ATCC 10031) and *Saccharomyces cerevisiae* (ATCC 9763). The most effective anti-fungal and anti-bacterial activities were recorded for the methanolic extract of intestines tract organs of *H. parva* against *S. cerevisiae* and *S. epidermidis* compared with other extracts (Mashjoor and Yousefzadi, 2017).

Some secondary metabolites also showed anti-bacterial activity. Lanosterol triterpenoids compound isolated from the sea cucumber *H. leucospilota* showed anti-bacterial activity against Gram-positive bacteria (*B. subtilis*, *Bacillus cereus* and *S. aureus*) (Nazemi et al., 2017). Also, two groups of steroidal and glycosides Saponins with Rf 0.9 and 0.05 were, respectively, identified from the sea cucumber *S. hermanni* and showed anti-bacterial activity (Salari et al., 2018).

The anti-fungal and anti-bacterial potential of the sea cucumber *H. leucospilota* collected from the Persian Gulf was investigated against *Aspergillus niger, C. albicans, S. aureus, P. aeruginosa* and *E. coli* by Mokhlesi et al. (2012). Methanol and water-methanol extracts that were prepared from Cuvierian organ and coelomic fluid, respectively, showed significant anti-fungal activities and had no inhibitory effect against bacteria (Mokhlesi et al., 2012). Also, methanol and chloroform extracts of the body wall, gonad and intestine of sea cucumber *H. leucospilota* collected from the Persian Gulf displayed anti-fungal activity against *A. niger* (Farjami et al., 2014a). Furthermore, the water-methanol extract from the sea cucumber *H. leucospilota* showed high anti-fungal activity against *A. niger* (Mohammadizadeh et al., 2013a). Likewise, *H. scabra* collected from the north coast of the Persian Gulf showed anti-fungal activity (Mohammadizadeh et al., 2013b); methanol and water-methanol extracts of the gonad, respiratory tree and body wall had anti-fungal activity against *A. niger* (Mohammadizadeh et al., 2013b). Furthermore, Sarhadizadeh et al. (2014) revealed that methanol extract, water-methanol extract and ethyl acetate extract from the gonads, respiratory tree, Cuvierian organs and body wall have anti-fungal activity against *A. niger* (Sarhadizadeh et al., 2014).

5.4.1.3 Anti-tumour properties

The anti-tumour activity of sea cucumber extracts plays a vital role in tumour development, which suggests that sea cucumber extracts may be promising therapeutic candidates following further confirming in vivo experiments and clinical trials. Salimi et al. (2017) noted that methanolic extract of the sea cucumber *H. parva* could act as an anti-cancer drug candidate by inducing cell death through a reactive oxygen species (ROS)-mediated mitochondrial pathway (Salimi et al., 2017). Also, Arast et al. (2018) demonstrated that the Persian Gulf sea cucumber *H. parva* extracts could selectively induce apoptosis in skin mitochondria isolated from melanoma cells (highly aggressive and deadly cancer cells) in mouse models. The ROS were increased in the skin mitochondria isolated from melanoma cells, using all the applied concentrations of sea cucumber extracts (i.e. 250, 500 and 1000 μg/mL) (Arast et al., 2018). In vitro findings demonstrated that mitochondrial ROS formation, mitochondrial membrane potential (MMP) collapse and mitochondrial swelling and cytochrome c release were significantly increased after the addition of different concentrations of *H. parva* only in cancerous mitochondria (Salimi et al., 2017). Consistently, different concentrations of *H. parva* significantly increased cytotoxicity and caspase-3 activations only in cancerous but not normal non-cancerous B lymphocytes (Salimi et al., 2017). Likewise, *H. parva* methanolic extract increased cytochrome c release, ROS formation, MMP and mitochondrial swelling. It decreased complex II activity in the mitochondria obtained from cancerous liver hepatocytes (Pourahmad et al., 2016; Seydi et al., 2015).

Assarian et al. (2012) noted that the coelomic fluid and aqueous extract of the sea cucumbers *Stichopus horrens* and *H. leucospilota* collected from the Persian Gulf around Khark Island have a good cytotoxic effect on K562 and Wehi-164 cancer cells (Assarian et al., 2012). Moreover, the methanolic extract of *H. parva* and *H. sabra* collected from the Persian Gulf has anti-cancer activity. Different concentrations (250, 500 and 1000 μg/mL) of the methanolic extract significantly induced ROS formation and mitochondrial swelling, decreased MMP disruption and increased cytochrome c release in hepatocellular and the melanoma cancer cell, in a time- and concentration-dependent manner (Razi et al., 2018a).

Sea cucumber extracts inhibit many types of cancer cells. The sea cucumber *H. leucospilota* extract as a novel resource of natural aquatic products can significantly inhibit cervical cancer cell growth (Baharara et al., 2016b). The synergistic effect of *H. leucospilota* organic extract with radiotherapy administrated cytotoxic effects through apoptosis induction on HeLa cancer cells (Baharara et al., 2016b). Furthermore, the survival of HeLa cell was decreased by increasing the concentration of extract of the sea cucumber *H. leucospilota*. Therefore, it has cytotoxic activity against cancer cell lines (Vaseghi et al., 2018). Moreover, different concentrations (250, 500 and 1000 μg/mL) of the n-hexane, diethyl ether and methanolic extracts of the sea cucumber *H. scabra* can potentially serve as an anti-cancer agent against hepatocellular carcinoma cancer cells (Razi et al., 2018b). Similarly, the ethyl acetate fraction of body wall extracts from *H. scabra* exhibited anti-cancer activity against human glioblastoma cells through the mitochondrial-mediated apoptotic pathway, and the compounds in the extract might be a novel candidate for anti-glioblastoma (Sangpairoj et al., 2016). Also, the water extract of sea cucumber *H. arenicola* has a remarkable anti-proliferative effect on CT26 tumour cells (Baharara et al., 2016a). Still, the Persian Gulf holothurians *H. scabra*, *H. parva* and *H. leucospilota* have anti-cancer potential. Based on cell lines, the more effective extracts were noticed for ethyl acetate fractions of Cuvierian tubule organs of *H. leucospilota* against human cancer cell line MCF-7 and ethyl acetate extracts of intestine tract organs of *H. parva* against HeLa cells

(Mashjoor and Yousefzadi, 2019). Additionally, the aqueous fraction of the sea cucumber *H. parva* showed the best effect among fractions tested against breast cancer cells MCF-7 (Ehsanpour et al., 2015).

Sea cucumbers may contain a variety of anti-tumour ingredients. Echinoside A, holothurin A and 24-dehydroechinoside A were isolated from the sea cucumber *H. leucospilota*, which exhibited moderate cytotoxic activity against the HeLa tumour cell line (Shushizadeh et al., 2019). Also, 80% ethanol fraction of saponin isolated from *H. leucospilota* offered promise an anti-cancer candidate; it demonstrated haemolytic activity. Additionally, the crude saponin extracted from the sea cucumber was cytotoxic to A549 cells (Soltani et al., 2014). Furthermore, a triterpene glycoside saponin was isolated and characterised from the sea cucumber *Holothuria atra* collected in the Persian Gulf together with the three known saponins 2—4. These compounds showed remarkably cytotoxic effect against HeLa cells, with IC50 values ranging from 1.2 to 2.5 µg/mL (Grauso et al., 2019). Moreover, the (Z)-2,3-diphenylacrylonitrile molecule, which is also known as α-cyanostilbene, was isolated from the methanolic extract of the sea cucumber *H. parva*. It increased ROS generation, the collapse of MMP, swelling in mitochondria and cytochrome c release in hepatocellular carcinoma liver mitochondria (Amidi et al., 2017).

5.4.1.4 Anti-viral properties

There is evidence that bioactive compounds of Persian sea cucumbers have anti-viral activity. From Bushehr port in the Persian Gulf, the cold water extract from the sea cucumber *Holothuria* sp. exhibited anti-viral activity against the KOS strain of herpes simplex virus type 1 in cell culture. The extract exhibited anti-viral activity not only against the virus adsorption to the cells but also on virus intracellular replication (Farshadpour et al., 2014). Also, the methanol extract of digestive organs and diethyl ether extract of the body wall of *H. leucospilota* have useful bioactive compounds. Diethyl ether extract obtained from the body wall demonstrated anti-viral properties and inhibited HIV-1. The digestive organs methanol extract, which had less cytotoxicity effect on the host cell, inhibited HIV-1 replication more significantly compared with other extracts (Bahroodi et al., 2018).

5.4.1.5 Anti-oxidant properties

The methanol and water-methanol extracts of the sea cucumbers *H. leucospilota* and *Bohadschia marmorata* collected from the Persian Gulf showed anti-oxidant activity effect (Permeh et al., 2013). Also, the extract of the sea cucumbers *H. leucospilota* and *S. hermanni* from the Persian Gulf showed anti-oxidant and α- glucosidase inhibitory effects. Regarding enzyme inhibition and anti-oxidant activity, the respiratory tree extract showed higher activity than methanol and dichloromethane extracts of gonads and body wall. In both sea cucumbers, the extract of the respiratory tree showed maximum inhibition of 34% and 40% for *S. hermanni* and *H. leucospilota*, respectively, on α-glucosidase activity (Abbasi et al., 2019). Furthermore, the body wall, intestines, gonads and respiratory tree of the sea cucumbers *H. scabra*, *H. parva* and *H. leucospilota* extracted by n-hexane, ethyl acetate and methanol have moderate anti-oxidant activity (Mashjoor et al., 2018).

5.4.1.6 Fertility improvement

Sea cucumbers have been recognised as a tonic remedy. The extract of the sea cucumber *H. arenicola* collected from the Persian Gulf has a protective effect against the detrimental effects of a low-frequency electromagnetic field in sperm parameters (Vasei et al., 2016). Also, the extracted saponin from the sea cucumber *H. leucospilota* can positively influence the maturation of pre-antral follicles by reducing oxidative stress, TNFα expression and inducing anti-oxidant enzyme activity in mice (Moghadam et al., 2016). Furthermore, the methanol extract of the sea cucumber *H. leucospilota* can improve oocyte maturation in the female mouse ovarian. Therefore, it could be suggested for use in the infertility modality (Khalilzadeh et al., 2016). Additionally, the Persian Gulf sea cucumber *H. leucospilota* extract influences the production of steroid hormones in molly fish (*Poecilia sphenops*). Injection of alcoholic extract of Persian Gulf sea cucumber (200 mg/kg) and non-polar extract of Persian Gulf sea cucumber (100 and 200 mg/kg) significantly diminished body testosterone and cholesterol levels in the molly fish (Golestani et al., 2016).

5.4.1.7 Other properties

In addition to the pharmacologic and therapeutic aspects described above, other studies revealed that sea cucumbers have other potential biological properties. For instance, sea cucumber extract was able to differentiate rat bone marrow mesenchymal stem cells into the osteogenic lineage. The appropriate dosage for the treatment of mesenchymal stem cells was determined to be less than 50 µg/mL. Also, oil red staining showed that sea cucumber alcoholic extract does not have the capacity of adipogenic induction. However, alizarin red staining and alkaline phosphatase revealed that a concentration of 25 µg/mL sea cucumber alcoholic extract was the most efficient concentration into osteogenic differentiation (Baharara et al., 2014). Additionally, injecting *H. arenicola* extract into the rats with diabetes results in a decrease and increase in glucose, insulin and β-amyloid, respectively (Sadoughi and Chamipa, 2016).

Furthermore, the sea cucumber (*H. leucospilota*) collected from Larak Island, Persian Gulf, can be considered a source of bioactive-producing actinobacteria that are potential candidates for anti-oxidant and cytotoxic compounds discovery (Gozari et al., 2018). Also, the sea cucumber *H. leucospilota* could be sources of anti-protozoal compounds, which could be used as a lead potent anti-leishmanial and cytotoxic agent; the methanol extract of the body wall, coelomic fluid, and Cuvierian organs of *H. leucospilota* showed an inhibitory effect against the *Leishmania* parasite (Khademvatan et al., 2016). Additionally, ethanol extract of the Persian Gulf sea cucumber *H. leucospilota* had a protective and therapeutic effect against carbon tetrachloride-induced hepatotoxicity in molly fish (Mohammadi et al., 2019). Moreover, the methanolic extract of *H. leucospilota* possesses a lethal effect on the parasite *Leishmania major* promastigotes and induces the apoptosis in parasites (Foroutan-Rad et al., 2016) (Table 5.4).

Table 5.4 Pharmacologic and medicinal activities of Persian sea cucumbers.

Species	Biological activities	References
Holothuria parva	Anti-microbial	Arast et al. (2018), Ebrahimi et al. (2018), Mashjoor and Yousefzadi (2017), Razi et al. (2018a), Salimi et al. (2017), Shadi and Oujifard (2019)
	Anti-tumour	Amidi et al. (2017), Ehsanpour et al. (2015), Pourahmad et al. (2016), Seydi et al. (2015)
	Anti-oxidant	Mashjoor et al. (2018).
Holothuria scabra	Anti-microbial	Mashjoor and Yousefzadi (2017), Mohammadizadeh et al. (2013b), Sangpairoj et al. (2016)
	Anti-tumour	Razi et al. (2018a), Sangpairoj et al. (2016)
	Anti-oxidant	Mashjoor et al. (2018)
Holothuria leucospilota	Anti-microbial	Adibpour et al. (2014), Farjami et al. (2014a, 2014b), Ghadiri et al. (2018), Kiani et al. (2014), Mashjoor and Yousefzadi (2017), Mohammadizadeh et al. (2013a), Mokhlesi et al. (2012), Nazemi et al. (2017)
	Anti-tumour	Assarian et al. (2012), Baharara et al. (2016a), Mashjoor and Yousefzadi (2019), Shushizadeh et al. (2019), Soltani et al. (2014), Vaseghi et al. (2018).
	Anti-viral	Bahroodi et al. (2018)
	Anti-oxidant	Mashjoor et al. (2018), Permeh et al. (2013)
	Improves fertility	Golestani et al. (2016), Khalilzadeh et al. (2016).
Stichopus hermanni	Anti-microbial	Salari et al. (2018), Sarhadizadeh et al. (2014)
	Anti-cancer	Abbasi et al. (2019)
	Anti-oxidant	Mashjoor et al. (2018)
Stichopus horrens	Anti-tumour	Assarian et al. (2012)
Holothuria atra	Anti-tumour	Grauso et al. (2019)
Holothuria arenicola	Anti-tumour	Baharara et al. (2016a)
	Improve fertility	Vasei et al. (2016)
Holothuria sp.	Anti-viral	Farshadpour et al. (2014)
Bohadschia marmorata	Anti-oxidant	Permeh et al. (2013)

5.5 Conclusion

Sea cucumbers aquaculture in the Persian Gulf is still in its early developmental stage, with some trials on the sea cucumbers *H. scabra* and *H. leucospilota*. The knowledge related to sea cucumber aquaculture (i.e. reproductive biology, habitat preference and ecology) in the Persian Gulf has not received much attention. However, Persian sea cucumbers have gained popularity among researchers in the Persian Gulf due to their nutritive value and therapeutic uses. Researchers concluded that sea cucumbers might be used in the future as an appropriate source of anti-microbial, anti-tumour, anti-viral, anti-oxidant and fertility improvement. These properties might vary according to the species, extracted organ and extraction method.

Furthermore, in-depth studies are required to isolate and identify the bioactive components. Persian sea cucumbers are a potential source for the discovery of natural anti-biotic compounds and drug development. Also, more investigations are needed to understand the mechanism of action of the bioactive compounds in the cells. Further characterisations are recommended for the detection of sea cucumber metabolites and clinical application.

References

Abbasi, H., Moein, S., Ehsanpoor, M., 2019. α-Glucosidase inhibition and antioxidant activities of respiratory tree, gonad, and body wall extracts of two species of sea cucumbers (*Holothuria leucospilota, Stichopus hermanni*) from Persian Gulf. Trends Pept. Protein Sci. 4, 10-11−6. https://doi.org/10.22037/tpps.v4i0.27517.

Adibpour, N., Nasr, F., Nematpour, F., Shakouri, A., Ameri, A., 2014. Antibacterial and antifungal activity of *Holothuria leucospilota* isolated from Persian Gulf and Oman Sea. Jundishapur J. Microbiol. 7. https://doi.org/10.5812/jjm.8708.

Afkhami, M., Ehsanpour, M., Forouzan, F., Bahri, A., 2012a. New observation of sea cucumber, *Holothuria* (Mertensiothuria) *hilla*, from Larak Island (Persian Gulf, Iran). Mar. Biodivers. Rec. 5, 1−3. https://doi.org/10.1017/S1755267212000425.

Afkhami, M., Ehsanpour, M., Khazaali, A., Dabbagh, A.R., Maziar, Y., 2012b. New observation of a sea cucumber, *Holothuria* (Thymiosycia) *impatiens*, from Larak Island (Persian Gulf, Iran). Mar. Biodivers. Rec. 5 https://doi.org/10.1017/S1755267212000425.

Afkhami, M., Ehsanpour, M., Nasrolahi, A., 2015. Two sea cucumber species (*Holothuria bacilli* and *H. insignis*): first record from the Persian Gulf. Mar. Biodivers. Rec. 8 https://doi.org/10.1017/S1755267215000718.

Afkhami, M., Ehsanpour, M., Nasrolahi, A., 2014. First record of a sea cucumber species (*Holothuria notabilis* Ludwing , 1875) from the Persian Gulf. Casp. J. Appl. Sci. Res. 3, 8−11.

Ahmed, M.I., 2009. Morphological, ecological and molecular examination of the sea cucumber species along the Red Sea coast of Egypt and Gulf of Aqaba: with the investigation of the possibility of using DNA barcoding technique as a standard method for sea cucumber ID (Doctoral dissertation, The University of Hull).

Ali-abadi, M.A.S., 2016. The study of population genetics structure of *Holothuria parva* in the Persian Gulf using mtDNA sequences. Int. J. Life Sci. Biotechnol. Pharma Res. 5. https://doi.org/10.18178/ijlbpr.5.1.14-17.

Amidi, S., Hashemi, Z., Motallebi, A., Nazemi, M., Farrokhpayam, H., Seydi, E., Pourahmad, J., 2017. Identification of (Z)-2,3-diphenylacrylonitrile as anti-cancer molecule in Persian Gulf sea cucumber *Holothuria parva*. Mar. Drugs 15. https://doi.org/10.3390/md15100314.

Arast, Y., Seyed Razi, N., Nazemi, M., Seydi, E., Pourahmad, J., 2018. Non-polar compounds of Persian Gulf sea cucumber *Holothuria parva* selectively induce toxicity on skin mitochondria isolated from animal model of melanoma. Cutan. Ocul. Toxicol. 37, 218–227. https://doi.org/10.1080/15569527.2017.1414227.

Assarian, M., Rakhshan, A., Khodadadi, A., Adibpour, N., Rezaee, S., 2012. Evaluation of cytotoxicity of two Persian Gulf sea cucumber species extracts on K562 and Wehi-164 cell lines and blood granulocyte cells. Res. Pharm. Sci. 7, 112.

Baharara, J., Amini, E., Afzali, M., Nikdel, N., Mostafapour, A., Kerachian, M.A., 2016a. Apoptosis inducing capacity of *Holothuria arenicola* in CT26 colon carcinoma cells in vitro and in vivo. Iran. J. Basic Med. Sci. 19, 358–365. https://doi.org/10.22038/ijbms.2016.6806.

Baharara, J., Amini, E., Namvar, F., Soltani, M., 2014. The effect of Persian Gulf sea cucumber alcoholic extract on osteogenic and adipodgenic differentiation of rat mesenchymal stem cells. JCT 5, 273–280.

Baharara, J., Amini, E., Vazifedan, V., 2016b. Concomitant use of sea cucumber organic extract and radiotherapy on proliferation and apoptosis of cervical (HeLa) cell line. Zahedan J. Res. Med. Sci. 6442. https://doi.org/10.17795/zjrms-6442.

Bahroodi, S., Nematollahi, M.A., Aghasadeghi, M.R., Nazemi, M., 2018. In vitro evaluation of the antiviral activity and cytotoxicity effect of *Holothuria leucospilota* sea cucumber extracts from the Persian Gulf. Infect. Epidemiol. Microbiol. 4, 153–157.

Bower, A.S., Hunt, H.D., Price, J.F., 2000. Character and dynamics of the red sea and Persian Gulf outflows. J. Geophys. Res. Ocean. 105, 6387–6414. https://doi.org/10.1029/1999jc900297.

Dabbagh, A.-R., Sedaghat, M.R., 2012. Breeding and rearing of the sea cucumber *Holothuria scabra* in Iran. SPC Beche-de-mer Inf. Bull. 49–52.

Dabbagh, A.-R., Sedaghat, M.R., Rameshi, H., Kamrani, E., 2011. Breeding and larval rearing of the sea cucumber *Holothuria leucospilota* Brandt (*Holothuria vegabunda* Selenka) from the northern Persian Gulf, Iran. SPC Beche-de-mer Inf. Bull. 35–38.

Dabbagh, A.R., Keshavarz, M., Mohammadikia, D., Afkhami, M., Nateghi, S., 2012a. *Holothuria scabra* (Holothuroidea: Aspidochirotida): first record of a highly valued sea cucumber, in the Persian Gulf, Iran. Mar. Biodivers. Rec. 5. https://doi.org/10.1017/S1755267212000620.

Dabbagh, A.R., Sedaghat, M.R., Keshavarz, M., Mohammadikia, D., 2012b. First record of a sclerodactylid sea cucumber, *Ohshimella ehrenbergii*(selenka, 1868) from Iranian waters (echinodermata: Holothuroidea: Sclerodactylidae). Indian J. Geo-Mar. Sci. Sci. 41, 117–120.

Ebrahimi, H., Vazirizadeh, A., Nabipour, I., Najafi, A., Tajbakhsh, S., Nafisi Bahabadi, M., 2018. In vitro study of antibacterial activities of ethanol, methanol and acetone extracts from sea cucumber *Holothuria parva*. Iran. J. Fish. Sci. 17, 542–551. https://doi.org/10.22092/IJFS.2018.116472.

Ehsanpour, Z., Archangi, B., Salimi, M., Salari, M., Zolgharnein, H., 2016. Morphological and molecular identification of *Holothuria* (Selenkothuria) *parva* from Bostaneh port, Persian Gulf. Indian J. Geo-Mar. Sci. 45, 405–409.

Ehsanpour, Z., Archangi, B., Salimi, M., Salari, M.A., Zolgharnein, H., 2015. Cytotoxic assessment of extracted fractions of sea cucumber *Holothuria parva* on cancer cell line (MCF7) and normal cells. J. Oceanogr. 6, 89–96.

Farjami, B., Nematollahi, M., Moradi, Y., Nazemi, M., 2014a. Derivation of extracts from Persian Gulf sea cucumber (*Holothuria leucospilota*) and assessment of its antifungal effect. Iran. J. Fish. Sci. 13 (4), 785–795. https://doi.org/10.22092/ijfs.2018.114396.

Farjami, B., Nematollahi, M.A., Noradi, Y., Irajian, G., Nazemi, M., 2014b. Study of antibacterial effect of the extracts of the sea cucumber (*Holothuria leucospilota*) of Persian Gulf on the *Escherichia coli*. Iran. J. Med. Microbiol.

Farshadpour, F., Gharibi, S., Taherzadeh, M., Amirinejad, R., Taherkhani, R., Habibian, A., Zandi, K., 2014. Antiviral activity of *Holothuria* sp. a sea cucumber against herpes simplex virus type 1 (HSV-1). Eur. Rev. Med. Pharmacol. Sci. 18, 333–337.

Fatemi, M.R., Mostafavi, P.G., Hamiz, Z., 2011. Identification of intertidal zone sea cucumbers (class: Holothuroidea) of Qeshm Island (Persian Gulf, Iran). J. Oceanogr. 2 (7), 57–65.

Foroutan-Rad, M., Khademvatan, S., Saki, J., Hashemitabar, M., 2016. *Holothuria leucospilota* extract induces apoptosis in Leishmania major promastigotes. Iran. J. Parasitol. 11, 339–349.

Ghadiri, M., Kazemi, S., Heidari, B., Rassa, M., 2018. Bioactivity of aqueous and organic extracts of sea cucumber *Holothuria leucospilota* (Brandt 1835) on pathogenic Candida and Streptococci. Int. Aquat. Res. 10, 31–43. https://doi.org/10.1007/s40071-017-0186-x.

Ghobadyan, F., Morovvati, H., Ghazvineh, L., Tavassolpour, E., 2012. An investigation of the macroscopic and microscopic characteristics of gonadal tubules in the sea cucumber *Holothuria leucospilota* (Persian Gulf, Iran). SPC Beche Mer Inf. Bull. 6–14.

Golestani, F., Naji, T., Sahaifi, H., 2016. Comparison effect of alcoholic and non-polar extract of Persian Gulf sea cucumber (*Holothuria leucospilota*) on steroid hormones levels in molly fish (*Poecilia sphenops*). J. Vet. Sci. Technol. 7, 385. https://doi.org/10.4172/2157-7579.

Gozari, M., Bahador, N., Jassbi, A.R., Mortazavi, M.S., Eftekhar, E., 2018. Antioxidant and cytotoxic activities of metabolites produced by a new marine *Streptomyces* sp. isolated from the sea cucumber *Holothuria leucospilota*. Iran. J. Fish. Sci. 17, 413–426. https://doi.org/10.22092/IJFS.2018.116076.

Grauso, L., Yegdaneh, A., Sharifi, M., Mangoni, A., Zolfaghari, B., Lanzotti, V., 2019. Molecular networking-based analysis of cytotoxic saponins from sea cucumber *Holothuria atra*. Mar. Drugs 17. https://doi.org/10.3390/md17020086.

Heding, S.G., 1940. Echinoderms of the Iranian Gulf. Holothuroidea. Danish Sci. Investig. Iran 2, 113–137.

Jadavi, N., Vaziri, S., Nabipour, I., 2015. Fat characteristics and fatty acid profile of sea cucumbers (*Holothuria scabra*) obtained from the coasts of the Bushehr province-Iran. Iran. South Med. J. 18, 992–1006.

Kämpf, J., Sadrinasab, M., 2006. The circulation of the Persian Gulf: a numerical study. Ocean Sci. 2, 27–41. https://doi.org/10.5194/os-2-27-2006.

Khademvatan, S., Eskandari, A., Saki, J., Foroutan-Rad, M., 2016. Cytotoxic activity of *Holothuria leucospilota* extract against Leishmania infantum in vitro. Adv. Pharmacol. Sci. 2016. https://doi.org/10.1155/2016/8195381.

Khalilzadeh, M., Baharara, J., Jalali, M., Namvar, F., Amini, E., 2016. The sea cucumber body wall extract promoted in vitro maturation of NMRI mice follicles at germinal vesicle stage. Zahedan J. Res. Med. Sci. 18. https://doi.org/10.17795/zjrms-3021.

Kiani, N., Heidari, B., Rassa, M., Kadkhodazadeh, M., Heidari, B., 2014. Antibacterial activity of the body wall extracts of sea cucumber (Invertebrata; Echinodermata) on infectious oral streptococci. J. Basic Clin. Physiol. Pharmacol. 25. https://doi.org/10.1515/jbcpp-2013-0010.

Majid, A., Maryam, E., Reza, D.A., Neda, S., Ghodrat, M., 2012. New observation of two sea cucumber species from Abu Musa Island (Persian Gulf, Iran). Eur. J. Exp. Biol. 2 (3), 611–615.

Maryam, E., Majid, A., Houshang, B.A., Abdolrasoul, D., 2012. First Report of a sea cucumber, Stichopus cf. monotuberculatus (Quoy & Gaimard, 1833), from Hengam Island (Persian Gulf, Iran). Eur. J. Exp. Biol. 2, 547–550.

Mashjoor, S., Yousefzadi, M., 2019. Cytotoxic effects of three Persian Gulf species of Holothurians. Iran. J. Vet. Res. 20, 19–26.

Mashjoor, S., Yousefzadi, M., 2017. Holothurians antifungal and antibacterial activity to human pathogens in the Persian Gulf. J. Mycol. Med. 27, 46–56. https://doi.org/10.1016/j.mycmed.2016.08.008.

Mashjoor, S., Yousefzadi, M., Pishevarzad, F., 2018. In vitro biological activities of holothurians edible sea cucumbers in the Persian Gulf. Indian J. Geo-Mar. Sci. 47, 1518–1526.

Mehrdost, M., Kamrani, E., Owfi, F., Ehsanpour, M., 2015. New observation of two species of sea cucumbers from the intertidal habitats of Hengam Island (Persian Gulf, Iran). Mar. Biodivers. Rec. 8.

Moghadam, F.D., Baharara, J., Balanezhad, S.Z., Jalali, M., Amini, E., 2016. Effect of Holothuria leucospilota extracted saponin on maturation of mice oocyte and granulosa cells. Res. Pharm. Sci. 11, 130–137.

Mohammadi, A., Naji, T., 2019. The protective and therapeutic effects of Persian Gulf sea cucumber (Holothuria leucospilota) on carbon tetrachloride-induced hepatotoxicity in Molly fish (Poecilia sphenops). Int. J. Ornam. Aquat. Res. 1, 2019.

Mohammadizadeh, F., Ehsanpor, M., Afkhami, M., Mokhlesi, A., Khazaali, A., Montazeri, S., 2013a. Antibacterial, antifungal and cytotoxic effects of a sea cucumber Holothuria leucospilota, from the north coast of the Persian Gulf. J. Mar. Biol. Assoc. U. K. 93, 1401–1405. https://doi.org/10.1017/S0025315412001889.

Mohammadizadeh, F., Ehsanpor, M., Afkhami, M., Mokhlesi, A., Khazaali, A., Montazeri, S., 2013b. Evaluation of antibacterial, antifungal and cytotoxic effects of Holothuria scabra from the North Coast of the Persian Gulf. J. Mycol. Med. 23, 225–229. https://doi.org/10.1016/j.mycmed.2013.08.002.

Mokhlesi, A., Saeidnia, S., Gohari, A.R., Shahverdi, A.R., Nasrolahi, A., Farahani, F., Khoshnood, R., Es'haghi, N., 2012. Biological activities of the sea cucumber Holothuria leucospilota. Asian J. Anim. Vet. Adv. 7, 243–249. https://doi.org/10.3923/ajava.2012.243.249.

Nazemi, M., Moradi, Y., Mortazavi, M.S., Mohebbi Nozar, S.L., Karimzadeh, R., Ghaffari, H., 2017. Isolation and identification of antibacterial compound from common name Holothuria leucospilota in the Persian Gulf. Iran. J. Fish. Sci. 19 (5), 2493–2500. https://doi.org/10.22092/ijfs.2019.118570.

Permeh, P., Gohari, A., Saeidnia, S., Jamili, S., 2013. Antioxidant activity of two sea cucumber species and brown algae from Persian Gulf. In: 2nd National Congress on Medicinal Plants.

Pourahmad, J., Motallebi, A., Seydi, E., Amidi, S., 2016. Selective toxicity of Persian Gulf sea cucumber (*Holothuria parva*) methanolic extract on liver mitochondria isolated from animal model of hepatocellular carcinoma. In: 3rd Int. Gastrointest. Cancer Congr..

Pourvali, N., Nabavi, M.B., Rezai, H., Doraghi, A.M., Mahvari, A.R., 2014. Shallow-water Holothuroidea (Echinodermata) from Hormuz Island in the Persian Gulf, Iran. World J. Fish Mar. Sci. 6, 395–399.

Rashidi, F.F., Kamrani, E., Ranjbar, M.S., 2017. Feeding regime of sea cucumber *Stichopus herrmanni* on coral reefs of southeast of the Qeshm Island, Persian Gulf. J. Oceanogr. 8, 35–41. https://doi.org/10.21833/ijaas.2018.01.014.

Razi, N.S., Arast, Y., Nazemi, M., Pourahmad, J., 2018a. The use of methanolic extract of Persian Gulf sea cucumber, Holothuria, as potential anti-cancer agents. Int. Pharm. Acta 2, 208–217. https://doi.org/10.22037/ipa.v1i2.21551.

Razi, N.S., Arast, Y., Nazemi, M., Pourahmad, J., 2018b. Selective toxicity of non-polar bioactive compounds of sea cucumber (*Holothuria scabra*) extracts on isolated mitochondria and hepatocytes of induced hepatocellular carcinoma rat model. Asian Pac. J. Cancer Biol. 3, 25–36. https://doi.org/10.31557/apjcb.2018.3.1.25-36.

Sadoughi, S.D., Chamipa, M., 2016. Effects of aqueous extract of *Holothuria arenicola* and low frequency electromagnetic field on serum insulin, glucose and beta-amyloid ($A\beta$1-42) in diabetic rats. KAUMS J. (FEYZ) 20, 1–10.

Salari-Aliabadi, M.A., Monjezi-Veysi, M., 2020. Application of calcareous spicules for the identification of sea cucumbers in the rocky shores of northern Persian Gulf. Indian J. Geo-Mar. Sci. 49, 281–286.

Salari, Z., Souinezad, I., Nazemi, M., Yousefzadi, M., 2018. Antibacterial activity of Saponin extracted from the sea cucumber (*Stichopus hermanni*) collected from the Persian Gulf. Iran. Sci. Fish. J. 27. https://doi.org/10.22092/ISFJ.2018.116406.

Salarzadeh, A., Afkhami, M., Bastami, K.D., Ehsanpour, M., Khazaali, A., Mokhleci, A., 2012. Proximate composition of two sea cucumber species *Holothuria pavra* and *Holothuria arenicola* in Persian Gulf. Ann. Biol. Res. 3, 1305–1311.

Salarzadeh, A., Afkhami, M., Ehsanpour, M., Mehvari, A., Darvish Bastami, K., 2013. Identification of sea cucumber species around Hengam Island (Persian Gulf, Iran). Mar. Biodivers. Rec. 6. https://doi.org/10.1017/S1755267212001212.

Salimi, A., Motallebi, A., Ayatollahi, M., Seydi, E., Mohseni, A.R., Nazemi, M., Pourahmad, J., 2017. Selective toxicity of Persian Gulf sea cucumber *Holothuria parva* on human chronic lymphocytic leukemia b lymphocytes by direct mitochondrial targeting. Environ. Toxicol. 32, 1158–1169. https://doi.org/10.1002/tox.22312.

Sangpairoj, K., Chaithirayanon, K., Vivithanaporn, P., Siangcham, T., Jattujan, P., Poomtong, T., Nobsathian, S., Sobhon, P., 2016. Extract of the sea cucumber, *Holothuria scabra*, induces apoptosis in human glioblastoma cell lines. Funct. Foods Health Dis. 6, 452. https://doi.org/10.31989/ffhd.v6i7.264.

Sarhadizadeh, N., Afkhami, M., Ehsanpour, M., 2014. Evaluation bioactivity of a sea cucumber, *Stichopus hermanni* from Persian Gulf. Eur. J. Exp. Biol. 4, 254–258.

Seydi, E., Motallebi, A., Dastbaz, M., Dehghan, S., Salimi, A., Nazemi, M., Pourahmad, J., 2015. Selective toxicity of Persian Gulf sea cucumber (*Holothuria parva*) and sponge (*Haliclona oculata*) methanolic extracts on liver mitochondria isolated from an animal model of hepatocellular carcinoma. Hepat. Mon. 15. https://doi.org/10.5812/hepatmon.33073.

Shadi, A., Oujifard, A., 2019. Antibacterial, cytotoxic and hemolytic activity of *Holothuria parva* sea cucumber from north Persian Gulf. Int. J. Environ. Sci. Technol. 16, 5937–5944. https://doi.org/10.1007/s13762-018-1956-8.

Shushizadeh, M., Mohammadi Pour, P., Mahdieh, M., Yegdaneh, A., 2019. Phytochemical analysis of *Holothuria leucospilota*, a sea cucumber from Persian Gulf. Res. Pharm. Sci. 14, 432–440. https://doi.org/10.4103/1735-5362.268204.

Soltani, M., Parivar, K., Baharara, J., Kerachian, M.A., Asili, J., 2014. Hemolytic and cytotoxic properties of saponin purified from *Holothuria leucospilota* sea cucumber. Rep. Biochem. Mol. Biol. 3, 43–50.

Soltani, M., Radkhah, K., Mortazavi, M.S., Gharibniya, M., 2010. Early development of the sea cucumber *Holothuria leucospilota*. Res. J. Anim. Sci. 4, 72–76. https://doi.org/10.3923/rjnasci.2010.72.76.

Tehranifard, A., 2011. Introducing a Holothorian sea cucumber species *Stichopus hermanni* form Kish island in the Persian Gulf in Iran. Int. Conf. Environ. Ind. Innov. IPCBEE 12, 138–143.

Tehranifard, A., Uryan, S., Vosoghi, G., 2006. Reproductive cycle of *Stichopus herrmanni* from Kish Island, Iran. Beche Mer Bull 24, 22–27.

Vaseghi, G., Hajakbari, F., Sajjadi, S., Dana, N., Ghasemi, A., Yegdaneh, A., 2018. Cytotoxic screening of marine organisms from Persian Gulf. Adv. Biomed. Res. 7.

Vasei, N., Baharara, J., ZAFAR, B., Amini, E., 2016. Evaluation the protective effect of sea cucumber extract (*Holothuria arenicola*) against injuries induced by electromagnetic field (50 Hz) on sperm parameters in male. J. Shahrekord Univ. Med. Sci. 18, 81–93.

Yahyavi, M., Afkhami, M., Javadi, A., Ehsanpour, M., Khazaali, A., Mokhlesi, A., 2012. Fatty acid composition in two sea cucumber species, *Holothuria scabra* and *Holothuria leucospilata* from Qeshm Island (Persian Gulf). Afr. J. Biotechnol. 11. https://doi.org/10.5897/ajb11.3529.

Sea cucumbers mariculture

6.1 Broodstock collection

Usually, mature adults are collected from nature by divers and then they are acclimated in the laboratory for breeding (Table 6.1). Larger-sized sea cucumbers could have greater reproductive output and expend higher reproductive energy than smaller ones (Chao et al., 1994). The adults of the sea cucumber *Holothuria poli* collected for breeding had a mean weight of 106.24 ± 9.47 g (n = 240; mean ± SE) (Rakaj et al., 2019). Also, Tolon and Engin (2019) collected *Holothuria tubulosa*

Table 6.1 Maturity size and spawning period for sea cucumber species.

Species	Size at sexual maturity	Spawning period	References
Holothuria leucospilota	180 g	Between November and March in the Great Barrier Reef (Australia). October to April in the Cook Islands. June and September in the Taiwan Province of China. February and May in Réunion.	Purcell et al. (2012)
Holothuria poli	135 mm	July to September in Oran Bay, Algeria, Mediterranean Sea.	Slimane-Tamacha et al. (2019)
		July to October in the Aegean Sea, the Mediterranean Sea.	Tolon and Engin (2019)
Stichopus horrens	16–18 cm	Can be artificially reproduced throughout the year except for May and July.	Hu et al. (2013), Purcell et al. (2012), Toral-Granda (2008)
Holothuria arguinensis	210–230 mm	Summer to autumn in the southern Iberian Peninsula.	Marquet et al. (2017)
Holothuria mammata	NA	Summer to autumn in the southern Iberian Peninsula.	Marquet et al. (2017)
Holothuria tubulosa	NA	July and September in the Alboran Sea, the Mediterranean Sea.	Ocaña and Sánchez (2005)

NA, *not available.*

Sea Cucumbers. https://doi.org/10.1016/B978-0-12-824377-0.00009-8

with a wet weight of 120.60 ± 19.56 g (n = 60) (Tolon and Engin, 2019). Likewise, adult specimens of *H. tubulosa* were collected with a mean weight of 247.3 ± 15.7 g (mean $\pm$ SE; 880 individuals) (Rakaj et al., 2018). For the sea cucumber *Stichopus horrens*, they were collected for spawning with a mean wet weight of >500 g (Hu et al., 2013). Additionally, Domínguez-Godino et al. (2015) collected 115 individuals as broodstock for the sea cucumber *Holothuria arguinensis*, and 25−30 individuals were used per spawning trial, with a mean weight of 279.3 ± 85.8 g (Domínguez-Godino et al., 2015). Moreover, Huang et al. (2018) collected *Holothuria leucospilota* sea cucumbers with a total weight of >400 g and a total length of >20 cm (Huang et al., 2018). Furthermore, adults of *Holothuria mammata* (n = 20) were collected with a mean weight of 182.1 ± 34.79 g (Domínguez-Godino and González-Wangüemert, 2018).

6.2 Gonadal development

The process of gonadal development is similar in the six sea cucumber species. Five stages of the development of the gonads can be observed in sea cucumber species (Conand, 1981; Ramofafia et al., 2000, 2001). These stages can be described as follows: pre-spawning or recovery, growing, early mature, mature and spent (Table 6.2) (Fig. 6.1) (Santos et al., 2017; Slimane-Tamacha et al., 2019).

Table 6.2 An example of gonad development in sea cucumbers. Macroscopic and microscopic characteristics of gonads in the sea cucumber *Holothuria leucospilota* from the Persian Gulf.

Maturity stage	Sex	Macroscopic	Microscopic
I Recovery	♀	The ovarian tubules are thin, white and transparent and very slightly branched and short. No oocytes empty tubules. Sex was undisguisable.	Tubules are empty with some previtellogenic oocytes lined the germinal epithelium.
	♂	Thick, short and yellow-white. Presence of early stage of sperm maturation.	Empty lumen area and presence of spermatocytes along the germinal epithelium.
II Growing stage	♀	White to light pink tubules. Presence of oocytes but the lumen is not entirely occupied. Sex can be distinguished.	Small oocytes and previtellogenic oocytes along germinal epithelium were observed.
	♂	Whitish colour tubules. The germinal epithelium extends towards the lumen of the tubule.	Tubules were seen with numerous infolds of the germinal epithelium with columns of spermatocytes along it and some spermatozoa in lumen area.

Table 6.2 An example of gonad development in sea cucumbers. Macroscopic and microscopic characteristics of gonads in the sea cucumber *Holothuria leucospilota* from the Persian Gulf.—*cont'd*

Maturity stage	Sex	Macroscopic	Microscopic
III Early mature	♀	Long, thick swollen and branched tubules, light pink to orange in colour.	Small oocytes were observed.
	♂	Long, creamy, branched and detailed tubules.	Tubules with a smooth wall were observed. Spermatocytes were significantly reduced, and the lumen was filled with spermatozoa.
IV Mature	♀	Maximum number of tubules with orange colour.	Mature oocytes and no small oocytes were present along the germinal epithelium.
	♂	Milky colour and thin walls.	Tubules' lumina were fully packed with spermatozoa and had fragile and smooth tubule walls. No proliferating zone with earlier stages along the germinal epithelium was seen.
V Spent stage	♀	Flaccid, regress, wrinkled and turned brown. Tubules are more or less empty.	Residual oocytes at various stages of deterioration were observed in the lumen. Some nutritive phagocytes begin to appear.
	♂	Tubules regress and become transparent. Tubules are flaccid and more or less empty.	Tubules in this stage showed elongated empty passages in the lumen following spawning and a few residual spermatozoa. No proliferating zone along the germinal epithelium containing spermatocytes was seen.

From Ghobadyan, F., Morovvati, H., Ghazvineh, L., Tavassolpour, E., 2012. An investigation of the macroscopic and microscopic characteristics\v of gonadal tubules in the sea cucumber Holothuria leucospilota (Persian Gulf, Iran). SPC Beche Mer Inf. Bull., 6—14.

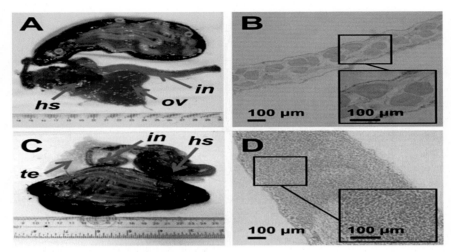

FIGURE 6.1

Macroscopic and microscopic gonad characteristics of the male and female *Holothuria leucospilota*. (A) Anatomy of the female specimen; ovary (ov). (B) Ovarian section. (C) Anatomy of the male specimen; intestine (in), haemal system (hs) and testis (te). (D) Testis section.

Credit: Huang, W., Huo, D., Yu, Z., Ren, C., Jiang, X., Luo, P., Chen, T., Hu, C., 2018. Spawning, larval development and juvenile growth of the tropical sea cucumber Holothuria leucospilota. *Aquaculture 488, 22–29. https://doi.org/10.1016/j.aquaculture.2018.01.013.*

6.3 Broodstock conditioning

It is crucial to maintain and acclimatize the broodstock in the laboratory before spawning (Duy, 2010). Adults are maintained in sterilised tanks (e.g. using chlorine) or ponds with at least 1 m depth and twice water exchange daily (morning and evening) (Hu et al., 2013). Huang et al. (2018) filtered and disinfected the water overnight using 20 ppm of trichloroisocyanuric acid with sufficient aeration. It was afterwards neutralised using 10 ppm of sodium thiosulphate with sufficient aeration until use. Also, the water quality of the tanks is maintained in optimum conditions. Generally, the oxygen is maintained between 4 and 6 mg/L, pH 8.10–8.30 and ammonia below 0.2 mg/L (Hu et al., 2013). Salinity together with the temperature should be maintained in optimum conditions for successful breeding. For instance, the sea cucumber *H. arguinensis* reduces its feeding, movements, absorption efficiency and growth when the seawater temperature drops below 19°C (Domínguez-Godino and González-Wangüemert, 2018). Furthermore, the researchers used several supplementary nutritions to the broodstock, which is summarised in Table 6.3.

The stocking density of the broodstock is reported. Domínguez-Godino and González-Wangüemert (2018) stocked *H. mammata* adults in 650-L indoor tanks at the

Table 6.3 Broodstock nutrition.

	Feeding	Temperature	References
Holothuria arguinensis	Sediment from nature environment	25–27°C	Domínguez-Godino et al. (2015)
Stichopus horrens	Sargassum thunbergii powder 10 g/m³ day	28°C	Hu et al. (2013)
Holothuria leucospilota	Live microalgae Amphora sp. and Chaetoceros muelleri	–	Huang et al. (2018)
Holothuria mammata	Broodstock was fed with dry powder of Zostera noltii based on 1.5% of the tank biomass once a week	25–27°C	Domínguez-Godino and González-Wangüemert (2018)
Holothuria poli	Natural sediment	24°C	Rakaj et al. (2019)
Holothuria tubulosa	Natural sediment and fish feed pellets were added as a supplement	23–24°C	Rakaj et al. (2018)

stocking density of 10 individuals/tank (Domínguez-Godino and González-Wangüemert, 2018). Also, Rakaj et al. (2018) stocked the broodstock of *H. tubulosa* at a density of about 40 animals per tank (30 L). The gonad index can be measured to assure optimum conditions for the broodstock (Huang et al., 2018):

$$\text{Gonad index}(\%) = (\text{Gonad wet weight} / \text{eviscerated wet weight (body wall)}) \times 100$$

6.4 Artificial induction of spawning

Gametogenesis in sea cucumbers is generally controlled by endogenous and seasonally variable environmental factors, which appear to interact in complex synergistic action. Several short-term environmental stresses are often applied to induce the sea cucumbers to spawn in captivity.

6.4.1 Thermal shock

The thermal shock can be carried out through transferring the broodstock to a tank with a higher or lower temperature (3–5°C) than the original one for about 1 h, with gentle stirring to keep the water homogeneous. After that, the water is replaced with new water of stable temperature. This method successfully induced spawning in the sea cucumber *H. poli* (Rakaj et al., 2019) and *H. arguinensis* (Domínguez-Godino et al., 2015). Also, *H. mammata* was effectively induced to spawn by thermal stimulation during the reproductive seasons (from July to October) (Domínguez-Godino and González-Wangüemert, 2018). However, this method did not efficiently induce spawning in the sea cucumbers *S. horrens* (Hu et al., 2013). Rakaj et al. (2018) successfully induced spawning in the sea cucumber *H. tubulosa* by combining the thermal shock and thermal stimulation; water temperature was gradually increased by 2—3°C for 2 days, and then, thermal shock was formed after 2 days by quickly raising the temperature by 3°C (Rakaj et al., 2018).

6.4.2 Gonadal stimulation

The gonad stimulation is carried out through extracting the sperm from the males and using it to initiate the spawning of the females. Although this method was successfully used for *Holothuria scabra* and *Cucumaria frondosa* (Hamel and Mercier, 1996; Agudo, 2006), it was unsuccessful with the sea cucumber *H. arguinensis* (Domínguez-Godino et al., 2015).

6.4.3 Drying

The drying method is conducted as follows: the sea cucumber adults are left in dry tanks or with 2 cm of water in tanks under the shade for about 1 h. Then, the water of steady temperature is added. This method was effectively induced spawning in *S. horrens* and *H. leucospilota* (Hu et al., 2013; Huang et al., 2018). Also, it was effectively used with the sea cucumber *H. scabra* (Agudo, 2006).

6.4.4 Mechanical shock

Mechanical shock is carried out through leaving the adults in a dry tank for 30 min. They are then exposed to severe water flow for about 30 min. Next, the adults are placed in tanks with normal water temperature. The sea cucumbers begin to spawn after about 1—1.5 h. Although this method was successfully used with the sea cucumbers *Holothuria atra* and *Bohadschia marmorata* (Laxminarayana, 2005), it was unsuccessful with the sea cucumber *H. poli* (Rakaj et al., 2019) and with the sea cucumber *H. tubulosa* (Rakaj et al., 2018).

6.4.5 Water pressure plus temperature

This method was reported as follows: *H. leucospilota* sea cucumbers are exposed to a powerful stream of water for 20 min. They are then subjected to an increase in the water temperature by 3—5°C for 1 h (Dabbagh et al., 2011).

6.4.6 Algae path

This method is conducted by adding dried *spirulina* (30 g every 300—500 L) or Algamac (0.1 g/L) for 1 h and then replacing the water with clean water with a stable temperature, and wastes are removed. Although this method was successfully used for *H. scabra* (Agudo, 2006), *Holothuria spinifera* and *Holothuria fuscogilva* (Asha and Muthiah, 2007; Battaglene et al., 2002), it was unsuccessful with the sea cucumbers *S. horrens* and *H. arguinensis*. Hu et al. (2013) added live microalgae (50,000—1,000,000 cells/mL), for example, *Dunaliella* sp. and *Chaetoceros muelleri*, in the tanks for a day. Also, Domínguez-Godino et al. (2015) used unicellular algae, such as *Tetraselmis chuii* (230,000,000 cells/250 L) and *Chaetoceros calcitrans* (880,000,000 cells/250 L). However, both trials did not successfully induce spawning.

6.5 Spawning behaviour

Sea cucumbers generate a distinctive posture when they release their gametes into the water. In most sea cucumber species, the attacking posture of the cobra was noticed when spawning occurs in the sea cucumber (Figs. 6.2—6.4). In addition to the cobra position, twisting, contractions and aligning to the walls of the tanks were commonly noticed before spawning in these six species. Similar spawning behaviour was also noticed in other sea cucumber species. In the sea cucumber *Pearsonothuria graeffei*, spawning behaviour was observed as follows: the animals

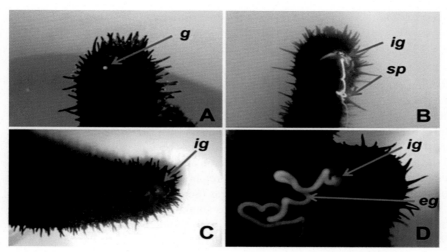

FIGURE 6.2

Spawning characterisation of *Holothuria leucospilota*. (A) Pre-spawning female; inflated gonopore (ig). (B) Ovulation; eggs (eg). (C) Pre-spawning male; gonopore (g). (D) Ejaculation; sperm (sp).

Credit: Huang, W., Huo, D., Yu, Z., Ren, C., Jiang, X., Luo, P., Chen, T., Hu, C., 2018. Spawning, larval development and juvenile growth of the tropical sea cucumber Holothuria leucospilota. *Aquaculture 488, 22—29. https://doi.org/10.1016/j.aquaculture.2018.01.013.*

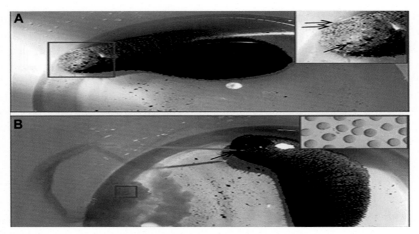

FIGURE 6.3

Spawning behaviour of the sea cucumber *Holothuria scabra* (A) head waving observed before spawning. Inset, higher magnification of sea cucumber head (red box), showing the location of gonopore (*single arrow*) and mouth (*double arrow*). (B) Sea cucumber spawning oocytes from gonopore. Inset, high magnification view of spawned oocytes (red box) indicating they were mature.

Credit: Chieu, H.D., Turner, L., Smith, M.K., Wang, T., Nocillado, J., Palma, P., Suwansaard, S., Elizur, A., Cummins, S.F., 2019. Aquaculture breeding enhancement: maturation and spawning in sea cucumbers using a recombinant relaxin-like gonad-stimulating peptide. Front. Genet. 10. https://doi.org/10.3389/fgene.2019. 00077. Licensed under CC BY 4.0.

stood up, waving the anterior part of the body slowly, and the gametes were spread into the water column (Purwati, 2003). Also, in the sandfish *H. scabra*, rolling movements, climbing the walls of the tanks, contractions of the body and lifting the front end of the body can be noticed before spawning (Agudo, 2006). In the sea cucumber *Apostichopus californicus*, the spawning posture has been described as being similar to the attacking posture of the cobra (Cameron and Fankboner, 1986). The sea cucumber *Cucumaria miniata* was observed stretching a third of the body in the water column, and the tentacles became motionless (Sewell and Levitan, 1992).

Elevation of the portion of the body was similarly noticed in the sea cucumber *Psolidium bullatum*, *C. miniata* (McEuen, 1988), *Holothuria mexicana* (Lawrence, 1982) and *H. atra* (Ramofafia et al., 1995). Furthermore, the spawning behaviour of the sea cucumber *Cucumaria lubrica* was observed in the laboratory and field as moving around, climbing the walls of the tanks or climbing the rocks in nature (McEuen, 1988). Moreover, the sea cucumber *C. frondosa* was observed before spawning, having incremental peristalsis of the body wall and extending the tentacles; the tentacles rapidly moved to help in extending the sperm in the males (Hamel and Mercier, 1996). Similar behaviour was noticed in *H. scabra*, *H. fuscogilva* and *Actinopyga mauritiana* (Battaglene et al., 2002).

Bohadschia vitiensis

FIGURE 6.4

The cobra posture during the spawning of the sea cucumber in nature.

Credit: Desurmont, A., 2005. Observations of natural spawning of Bohadschia vitiensis and Holothuria scabra versicolor. SPC Beche-de-mer Inf. Bull. 98848.

6.6 Fertilisation

Fertilisation is the union of an egg and a sperm (Fig. 6.5). Sea cucumbers mostly induced to spawn during the peak reproductive season. Usually, the males release the sperms first, and then the females are induced to release oocytes in the water column (McEuen, 1988). Thus, males and females can be distinguished from each other. Once the spawning occurs, the males can be removed to another tank (Rakaj et al., 2019).

When spawning sea cucumbers, UV-sterilised water should be provided in the tanks. The males release the sperm first from the gonopore on the head (Fig. 6.6). The turbidity of the tanks should be monitored by removing some males or carefully replacing the water, which can keep the sperm intensity at a moderate level. After 10−15 min, the males should be removed from the tanks to avoid polyspermy (i.e. an egg that has been fertilised by more than one sperm). To count the eggs, the beaker with eggs should be stirred gently, and then, 3−5 subsamples (1-mL) are collected and counted under a microscope (Dabbagh et al., 2011). The eggs

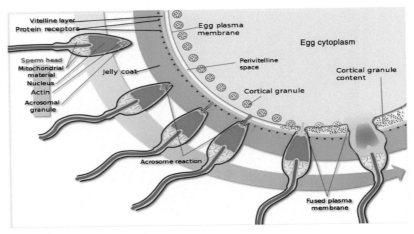

FIGURE 6.5

The acrosome reaction for a sea urchin; a similar process.

Credit: Public domain via Wikimedia Commons.

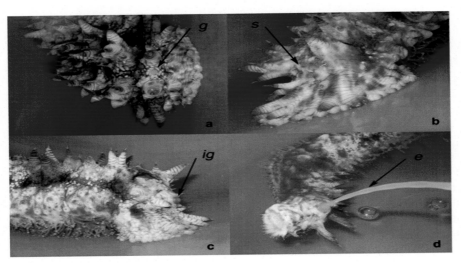

FIGURE 6.6

Spawning of *Stichopus horrens*. (A) Pre-spawning of the male; gonopore (g).(B) Ejaculation; sperm (s). (C) Pre-spawning of the female; inflated gonopore (ig). (D) Ovulation; egg (e).

Credit: Hu, C., Li, H., Xia, J., Zhang, L., Luo, P., Fan, S., Peng, P., Yang, H., Wen, J., 2013. Spawning, larval development and juvenile growth of the sea cucumber Stichopus horrens. *Aquaculture 404–405, 47–54. https://doi.org/10.1016/j.aquaculture.2013.04.007.*

Table 6.4 Mature egg diameter for sea cucumber species.

Sea cucumber species	Mature egg diameter (mean ± SE)
Holothuria leucospilota	140.80 ± 0.84 μm
Stichopus horrens	114 ± 2.8 μm
Holothuria arguinensis	108.30 ± 5.57 μm
Holothuria poli	203.7 ± 10.21 μm
Holothuria tubulosa	151.2 ± 2.1 μm

From Domínguez-Godino, J.A., Slater, M.J., Hannon, C., González-Wangüermert, M., 2015. A new species for sea cucumber ranching and aquaculture: breeding and rearing of Holothuria arguinensis. Aquaculture 438, 122—128. https://doi.org/10.1016/j.aquaculture.2015.01.004; Domínguez-Godino, J.A., González-Wangüemert, M., 2018. Breeding and larval development of Holothuria mammata, a new target species for aquaculture. Aquacult. Res. 49, 1430—1440. https://doi.org/10.1111/are. 13597; Hu, C., Li, H., Xia, J., Zhang, L., Luo, P., Fan, S., Peng, P., Yang, H., Wen, J., 2013. Spawning, larval development and juvenile growth of the sea cucumber Stichopus horrens. Aquaculture 404—405, 47—54. https://doi.org/10.1016/j.aquaculture.2013.04.007; Huang, W., Huo, D., Yu, Z., Ren, C., Jiang, X., Luo, P., Chen, T., Hu, C., 2018. Spawning, larval development and juvenile growth of the tropical sea cucumber Holothuria leucospilota. Aquaculture 488, 22—29. https://doi.org/10.1016/j. aquaculture.2018.01.013; Rakaj, A., Fianchini, A., Boncagni, P., Lovatelli, A., Scardi, M., Cataudella, S., 2018. Spawning and rearing of Holothuria tubulosa: a new candidate for aquaculture in the Mediterranean region. Aquacult. Res. 49, 557—568. https://doi.org/10.1111/are.13487; Rakaj, A., Fianchini, A., Boncagni, P., Scardi, M., Cataudella, S., 2019. Artificial reproduction of Holothuria polii: a new candidate for aquaculture. Aquaculture 498, 444—453. https://doi.org/10.1016/j.aquaculture. 2018.08.060.

should be washed several times by clean seawater in order to remove excess sperm, and females should not be disturbed during spawning (Domínguez-Godino and González-Wangüemert, 2018).

Among four different stocking densities (1, 5, 15 and 30 eggs/mL) tested, the optimal stocking density for the highest hatching and survival rates of the fertilised eggs of *H. tubulosa* was determined as 1—5 eggs/mL (Günay et al., 2018). The fertilised eggs should be examined under a microscope and can be removed by siphoning into a sieve from the tanks. After fertilisation, the first polar body can be noticed in the spherical eggs (Laxminarayana, 2005) (Table 6.4).

6.7 Embryo, larval and juvenile development

After fertilisation (about 1 h), the development of the embryo should be tracked continually (Rakaj et al., 2019). The developmental biology of the six species is very similar. All develop into a planktotrophic larva, called the auricularia, and it metamorphoses through a doliolaria stage into a pentacula.

Rearing the pelagic stages requires constant monitoring of the culture units (Table 6.5). The larvae can be transferred to a sterilised tank (using, e.g. chlorine or sodium hypochlorite) that is filled with filtered and UV-sterilised water, ambient temperature and salinity, sufficient aeration and gentle water circulation. Deformed or dead larvae, faeces from larvae and excess food should be removed at regular

Table 6.5 Water quality parameters during the larvae stage.

Species	Temperature	Salinity (ppt)	pH	Water exchange	References
Holothuria arguinensis	27—28°C	n.a.	n.a.	40%, every day	Domínguez-Godino et al. (2015)
Stichopus horrens	30°C	n.a.	7.9—8.2	n.a.	Hu et al. (2013)
Holothuria mammata	25—28°C	n.a.	n.a.	50% volume daily	Domínguez-Godino and González-Wangüemert, 2018
Holothuria tubulosa	24°C	n.a.	n.a.	Water was partially exchanged (50%) using a 60-µm mesh screened siphon	Rakaj et al. (2018)
Holothuria poli	24°C	36 and 37	n.a.	n.a.	Rakaj et al. (2019)
S. horrens	25—27°C	n.a.	7.9—8.2	n.a.	Huang et al. (2018)
Holothuria leucospilota	29—33°C	27—30	7.9—8.2	50% of the bottom seawater with filter (with diameter of 75 µm) for one time, twice a day	Huang et al. (2018)

n.a., *not available.*

intervals by gently siphoning the tank base. Ethylenediaminetetraacetic acid (2 mg/L) can be added from the second day up until the pentactula stage to improve water quality (Rakaj et al., 2018). The copepods can harm the larvae, especially <5 mm, by competing for food, harming their body and praying on them. The copepods and ciliates can be controlled by adding Dipterex (1—3 ppm for 1—3 h), which is enough to kill the copepods within 2—3 h, and then, water change should be followed later (Hu et al., 2013).

When the embryo has a functional gut, the larvae can be fed twice a day on a mix of the microalgae. The functional gut appeared 43 h after fertilisation in the sea cucumber *S. horrens* (Hu et al., 2013) and about 30—36 h after fertilisation in the sea cucumber *H. leucospilota* (Huang et al., 2018). Also, in *H. tubulosa* and *H poli,* the gut appeared on the third day following fertilisation (Rakaj et al, 2018,

2019). The stomach of the larvae should be observed under the microscope to detect the green algae, if not the feed can be increased. *T. chuii* and *C. calcitrans* were introduced to the sea cucumber *H. arguinensis* larvae at gradual increasing concentrations of 20,000—40,000 cells/mL (Domínguez-Godino et al., 2015). For the sea cucumber *S. horrens*, a mixture of microalgae consisting of *C. muelleri*, *Dunaliella* sp. and *Chlorella pyrenoidosa* (7:2:1) was provided as food for the larvae at gradual increasing concentrations of 10,000—150,000 cells/mL in the early stages, reaching 35,000—45,000 cells/mL (Hu et al., 2013). Also, a mixture of *C. muelleri* live microalgae (10,000—30,000 cells/mL) and *Saccharomyces cerevisiae* yeast powder (0.5—1.0 g per tonne seawater) was introduced to *H. leucospilota* larvae (Huang et al., 2018). Furthermore, the microalgae *T. chuii* and *C. calcitrans* (1:1) were introduced to *H. mammata* larvae at gradually increasing concentrations of 20,000 cells/mL to 40,000 cells/mL (Domínguez-Godino and González-Wangüemert, 2018). Likewise, the microalgae *Isochrysis galbana* and *C. calcitrans* (1:1) were introduced to *H. poli* larvae (Rakaj et al., 2019). Among three feeding regimes tested for *H. poli* larvae, the highest feeding regime of 20,000—40,000 cells/mL resulted in the highest metamorphosis of larvae into juveniles (Rakaj et al., 2019). However, the highest feeding regime prevented the metamorphosis of *H. tubulosa* larvae, and the lowest feeding concentration of 5000—10,000 cells/mL resulted in high numbers of metamorphosed larvae (Rakaj et al., 2018) (Fig. 6.7).

The auricularia larvae metamorphose into the swimming doliolaria larvae. Once the doliolaria appears, the settlement substrates are added with feed supplementation (Rakaj et al, 2018, 2019). The larvae need the functional primary tentacles to settle. Hu et al. (2013) provided the settlement substrates at the late auricularia stage when 5% of *S. horrens* larvae metamorphosed into doliolaria. Also, Huang et al. (2018) observed the primary tentacle primordium in the late auricularia stage beside the stomach of the *S. horrens* larvae.

The settlement substratum can be made from various materials (Figs. 6.8 and 6.9). The settlement substrate can be immersed first in cultures of diatoms for a few days to promote biofilm development, which will induce the larvae to settle on the substrate. Diatom biofilm of *Navicula* sp., *Nitzschia* sp. and *Phaeodactylum tricornutum* was used for the settlement of the sea cucumbers *H. poli* and *H. tubulosa* (Rakaj et al, 2018, 2019). Also, the microalgae of *Amphora* sp. were cultured in the settlement plates of the sea cucumber *H. leucospilota* (Huang et al., 2018).

The doliolaria larvae metamorphosis into pentactula larvae in their way to metamorphosis from a pelagic into a benthic life. Pentactula larvae can be observed for the first time using their tentacles to attach the substrate. As they settle down, adding diatoms and fresh or dried algae is essential as an additional source of food. Algamac 3050 was added at a concentration of 0.25 g/m^3 for the cultured larvae (Domínguez-Godino et al., 2015; Domínguez-Godino and González-Wangüemert, 2018; Rakaj et al, 2018, 2019). Hu et al. (2013) fed the early juveniles twice a day using a mixture of 1:1:2:5 of oceanic red yeast: microalgae powder *Dunaliella* sp.: powder of *spirulina* sp.: algae powder of *Sargassum thunbergii*, respectively. It was merged in the

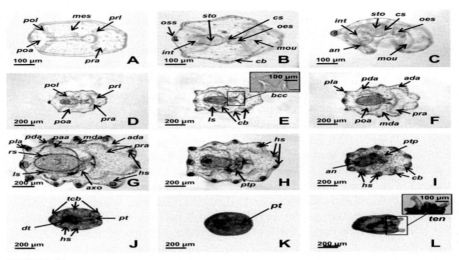

FIGURE 6.7

Larval development of the sea cucumber *Holothuria leucospilota*. (A–C): Early auricularia; preoral loop (prl), postoral loop (pol), postoral arm (poa), preoral arm (pra), mesenchyme (mes), ossicle (oss), intestine (int), stomach (sto), cardiac sphincter (cs), oesophagus (oes), mouth (mou), anus (an). (D–E): Mid-auricularia; buccal ciliated cavity (bcc), axohydrocoel (axo), left somatocoel (ls). (F–G): Auricularia; posterior-lateral arm (pla), posterior-dorsal arm (pda), anterior-dorsal arm (ada), mid-dorsal arm (mda), right somatocoel (rs), hyaline sphere (hs). (H): Late auricularia; primary tentacles primordium (ptp). (I–K): Doliolaria; transverse ciliary band (tcb), primary tentacle (pt), digestive tract (dt). L: Pentactula; tentacle (ten).

Credit: Huang, W., Huo, D., Yu, Z., Ren, C., Jiang, X., Luo, P., Chen, T., Hu, C., 2018. Spawning, larval development and juvenile growth of the tropical sea cucumber Holothuria leucospilota. *Aquaculture 488, 22–29. https://doi.org/10.1016/j.aquaculture.2018.01.013.*

seawater for 1 h and then was separated by a sieve (60 μm) (Hu et al., 2013). Huang et al. (2018) fed the *H. leucospilota* juvenile three times a day: in the morning, 1–5 L of live microalgae (density of 50,000–100,000 cells/mL) was added into 1000 L of cultured seawater; in the afternoon, 1–3 g of *Spirulina* sp. algae powder was added into 1000 L of cultured seawater; at dusk, 1–3 g of *Chlorella* sp. algae powder was added into 1000 L of cultured seawater. Furthermore, high survival rates and larger larval and stomach sizes were noticed when fed *H. arguinensis* larvae with *C. calcitrans*: *T. chuii*: *I. galbana* (1:1:1) than the ones fed with the other diets (Domínguez-Godino and González-Wangüemert, 2019a).

Once the juveniles appear Fig. 6.10, they can be bred to reach an appropriate size, which will help to reduce the possibility of predation (post settled juvenile) to increase the survival rate Fig. 6.11. The high mortality of *S. horrens* juvenile was noticed and lasted about 2 weeks from the day 34–46 after settlement, likely

FIGURE 6.8

Polyethene sheets as substrates for sea cucumbers.

Credit: Mohamed Mohsen.

FIGURE 6.9

Corrugated polyethene plates.

Credit: Mohamed Mohsen.

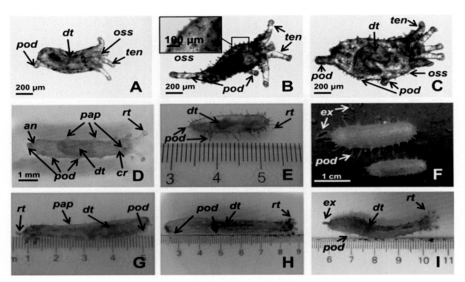

FIGURE 6.10

Juvenile development of *Holothuria leucospilota*. (A) 30-day juvenile; podia (pod), ossicle (oss), digestive tract (dt), tentacle (ten). (B) 34-day juvenile. (C) 37-day juvenile. (D) 55-day juvenile. (E) 79-day juvenile. (F) 79-day juveniles in the bottom of the black tank; excreta of the juvenile sea cucumbers (ex). (G) 120-day juvenile; ramified tentacle (rt), calcareous ring (cr), papillae (pap), anus (an). (H) Juvenile sea cucumber after 120 days. (I) The moving and excreting juvenile sea cucumber after 120 days.

Credit: Huang, W., Huo, D., Yu, Z., Ren, C., Jiang, X., Luo, P., Chen, T., Hu, C., 2018. Spawning, larval development and juvenile growth of the tropical sea cucumber Holothuria leucospilota. *Aquaculture 488, 22–29. https://doi.org/10.1016/j.aquaculture.2018.01.013.*

FIGURE 6.11

Juvenile rearing in cages inside ponds.

Credit: Obtained by Mohamed Mohsen.

because of the predator copepods. After the insecticide cleared most of the copepods, the mortality rate was decreased (Hu et al., 2013). After 80 d, Hu et al. (2013) compared the growth of juvenile *S. horrens* in tanks and cages. The survival rate of the juveniles in the indoor tanks was higher than in the cages, possibly because of the lack of food in the cages. Also, cage abrasions in the seabed might allow juveniles to escape from the cage through flaws or holes.

6.8 Environmental factors

Stress reflects a state of disturbance of animal welfare that is induced by a physiological or physical stressor, which challenges the homoeostatic power of an organism and/or threatens its survival (Colombo et al., 1989). In sea cucumbers, stress can lead to disease infection. Prevention of disease-causing factors is critical for the successful farming of sea cucumbers. Along with the rapid expansion and intensification of sea cucumber farming, various diseases have been discovered, causing economic losses and becoming a barrier to the sustainable development of sea cucumber farming. The most widespread pathogen reported from cultured sea cucumbers is bacteria. Skin ulceration bacterial disease was noticed in the sea cucumber *H. arguinensis* as a result of abiotic stress (Cánovas et al., 2019). The best way to avoid such bacterial disease is to regularly check the water quality parameters and the health of sea cucumbers and to eliminate sick individuals.

Water temperature regulates the major physiological processes of sea cucumbers. For instance, the feeding rate of the sea cucumber *H. tubulosa* increases as water temperature increases; the individuals ingest more sediment in summer than in winter (Coulon and Jangoux, 1993). Also, Günay et al. (2020) reported that the main digestive enzymes (i.e. protease, lipase and amylase) are strictly dependent on water temperature (Günay et al., 2020). Additionally, respiratory rates increase with increasing seawater temperature. Oxygen consumption of an average individual (7 g dw body wall) ranges from 0.409 (14°C) to 1.300 (26°C) mg O_2/h (Coulon et al., 1992). The drop in water temperature below the optimum temperature for the sea cucumber species can negatively affect the growth rate of the sea cucumber. Günay et al. (2015) reported that optimum growth temperature for *H. tubulosa* was between 20 and 25°C for the samples collected from the sea with the mean weight of 20.48 ± 2.33 g. Little or no growth was seen at 15°C and 30°C depending on the hibernation and aestivation (Günay et al., 2015). Also, juveniles of *H. poli* decreased feeding and lost their weight at low-temperature conditions (15°C) (Tolon, 2017).

Almost all Holothuroidea are stenohaline; therefore, it is important to avoid sea cucumber aquaculture in areas subjected to freshwater injection. *H. tubulosa* juveniles are unsuitable for low and high salinities. The *H. tubulosa* juveniles at 25 ppt salinity started to die on the fifth day at both 15°C and 25°C temperature trials. Also, *H. tubulosa* had negative specific growth rates (−0.81 and −0.55%/day) at 30 and 45 ppt in summer and winter temperature conditions. The optimum salinity for the sea cucumber *H. tubulosa* is 38 ppt in the summer and winter. However, the sea cucumber *H. poli* can be acclimated to high salinities (45 ppt), but with limited tolerance to the low salinities. The optimum salinity for the sea cucumber *H. poli* is 38 ppt, and *H. poli* showed a similar growth rate at 45 ppt (Tolon, 2017). Furthermore, Yu et al. (2013) reported that *H. leucospilota* might have a wide tolerance of salinity variation.

The stocking density is a critical factor during larval rearing. Lower stocking density results in higher survival rates. *H. arguinensis* juveniles showed the best growth at a stocking density of 5 ind./m^2, decreasing significantly as the stocking density increased (Domínguez-Godino and González-Wangüemert, 2018). Also, a density of 6 ind/m^2 is recommended for stocking juvenile *H. tubulosa* under rearing conditions. Stocking density of 15 ind/m^2 is not favourable for the long term as the growth rate is negligible, where 30 ind/m^2 should be avoided in tank-based rearing units (Tolon et al., 2017a).

Some studies were conducted to provide better environmental conditions for the juveniles. Sea cucumber *H. tubulosa* juveniles (35.02 ± 0.52 g) had the best growth performance with 1 mm sediment size (Tolon et al., 2015). During the nursery rearing of the juveniles, it is not recommended to use macrophyte detritus as supplemental food in the sediment because it can increase anoxic sediment conditions. Instead, the authors encourage using artificial materials in the nursery system to increase the surface area for biofilm growth and provide shelter for the light-sensitive juveniles (Palomar-Abesamis et al., 2018a). Furthermore, wild and hatchery-reared juveniles are most active at night. However, deviations in behaviour of hatchery-reared juveniles under laboratory conditions indicate some degree of acclimation to an artificial environment with minimal threats and a decreased sensitivity to light (Palomar-Abesamis et al., 2018b).

6.9 Grow-out methods

Farming sea cucumbers was firstly attempted with the sea cucumber *Apostichopus japonicus*. Later, its farming methods were adopted for the sea cucumber *H. scabra* and several sea cucumber species. Olaya-Restrepo et al. (2018) observed that the sea cucumber *H. arguinensis* achieves asymptotic size quickly, even faster than tropical commercially valuable sea cucumbers, such as the lolly fish, *Thelenota ananas*, *Stichopus chloronotus* or *Isostichopus fuscus*, which increase its potential for aquaculture (Olaya-Restrepo et al., 2018). Also, Domínguez-Godino and González-Wangüemert (2019b) reported that the sea cucumber *H. arguinensis* adults could be maintained under tank conditions, promoting their growth. They showed an increase of specific growth rate 0.2%/d with a mean organic matter content of 90.07 ± 11.5 mg/g in the offered sediment (Domínguez-Godino and González-Wangüemert, 2019b). Furthermore, Li et al. (2013) reported that juveniles and adults of the sea cucumber *S. horrens* are suitable for cage culture with adequate size and area (Li et al., 2013).

6.9.1 Earthen ponds

Several techniques are used for growing the sea cucumbers A. japonicus and H. scabra to the market size, which might be likewise adopted to other sea cucumber species. The sea cucumber H. scabra, for instance, is released in ponds with 10 g (about 6 cm), preferably >10 cm, reducing the possibility of predation. The seedlings are acclimated in the ponds before releasing them. The bottom of the ponds is adapted to suit the lifestyle of sea cucumbers. The lime powder is added $(20-30 \text{ g/m}^2)$ for $1-2$ d as a tool for disinfection. The sediment can be enriched with animal dust that is mixed with paddy bran in an equal ratio at a rate of $0.2-0.5 \text{ kg/m}^2$ (Agudo, 2006). Additional feed is rarely added, including sea mud with fish powder. The stocking density of sea cucumbers is a crucial factor for optimum growth rates. The sea cucumber H. scabra can be stocked in ponds at a density of $1.5-3$ tonnes/ha without additional feed. The juveniles reached from 30 to 300 g in 3 months. However, growth slowed down when densities exceeded $150-300 \text{ g/m}^3$ (Pitt and Duy, 2004). Also, stocking density of 20 ind./100 L is ideal for A. japonicus (5.12 ± 0.09 g), with a suitable food source (Dong et al., 2010).

6.9.2 Sea pens

Sea pens system is a cagelike rectangular or squared net fixed by wood in its angles. It can be a semi-intensive form of a small-scale aquaculture system because they offer great control over initial site selection (Purcell et al., 2012). Farmers constructed pens of $225-600 \text{ m}^2$ in nearshore seagrass beds using locally available material, such as nylon fishing nets and wood. The stocking density is reported as $200-250 \text{ g/m}^2$ in New Caledonia and $100-771 \text{ g/m}^2$ in Madagascar for H. scabra (Lavitra et al., 2010; Purcell et al., 2012; Purcell and Simutoga, 2008; Robinson and Pascal, 2012).

In Southwestern Madagascar, hatchery-reared H. scabra juveniles (15 g) were stocked in sea pens with 625 and 900 m^2 at a density of 0.5 ind/m^2. Survival rates from the initial trials ranged from 35% to 80%. Sea cucumbers reached a minimum of 300 g in $4-12$ months. The growth rates ranged from 1.0 to 1.8 g/d. The survival rates were increased by releasing the juveniles into protected nursery enclosures until they reach about 50 g, so they will be able to withstand predation (Robinson and Pascal, 2012).

6.9.3 Suspended culture

Various models of suspended culture are used in China, and the cage culture system is profitable for A. japonicus. These models are made from plastic with a top open or mesh with a zipper. The cages are suspended nearest to the surface or on the bottom, which provides protection and a source for growing the natural food. Additional food can be added once a week, or another species can be added with sea cucumbers as a food source for growing sea cucumbers much better. In this case, sea cucumber can be added on the bottom plate (Mu and Song, 2005; Yuan et al., 2008).

6.9.4 Co-culture

Co-culture of the sea cucumber with other species has received much interest due to their feeding strategy and their position in the food chain. Sea cucumbers are potential candidates for the integrated multi-trophic aquaculture (IMTA) systems or the rotational cultured systems. IMTA is defined as a balanced system that provides the by-products of one cultured species to another, which can achieve high income and environmental remediation. The rotational cultured system is farming one species after harvesting the other.

Sea cucumbers have gained much interest as a critical component of IMTA. Such a system should provide an opportunity for waste bioremediation and an increase in economic return if managed at the optimal growing condition for each co-cultured species. For instance, caution should be paid when co-culturing sea cucumbers with shrimp. Co-culture of the sea cucumber with shrimp has been tested, which gained a non-satisfied result due to the predation of the shrimp on the sea cucumber (Pitt et al., 2004). This predation was avoided by using larger juveniles of the sea cucumbers than of the shrimp (Li et al., 2014). Also, controlling the water quality is crucial, mainly when farming shrimp and sea cucumber in the earthen ponds (Purcell et al., 2006).

Furthermore, controlling the disease is crucial when using a rotational culture system (Zamora et al., 2018). Moreover, it is recommended to determine the suitable density of the sea cucumber when establishing the co-culture system as the density has a significant effect on the growth rate (Yokoyama, 2013; Zamora et al., 2018). Additionally, when integrating sea cucumber in the suspended cages at a subtropical fish farm, it is recommended to put the sea cucumber inside the suspended cages rather than beneath the suspended cages, which can lead to a better survival rate for the sea cucumber *H. leucospilota* (Yu et al., 2012b).

Co-culturing the sea cucumber with different marine species has been tested, mainly using the two main cultured species of *H. scabra* and *A. japonicus*. Sea cucumber was successfully co-cultured with abalone (Kim et al., 2015), fish (Tolon et al., 2017b), shrimp (Zhou et al., 2017; Li et al., 2014), mussel (Slater and Carton, 2007), scallop (Ren et al., 2012), seaweed (Namukose et al., 2016) and sea urchins (Wang et al., 2008). Furthermore, beneficial co-culture of the sea cucumber *H. leucospilota* has been observed with *Sargassum hemiphyllum* (Peng et al., 2012) and with shrimp *Litopenaeus vannamei* (Yu et al., 2012a). Also, sea cucumbers were successfully integrated under or in net cages (Kang et al., 2003; Ren et al., 2012; Namukose et al., 2016), in ponds (Wang et al., 2007, 2008; Zhou et al., 2017) and integrated into the recirculated aquaculture system (MacDonald et al., 2013) (Table 6.6).

Table 6.6 Farming models of sea cucumber with different species.

IMTA system	Species	Density	References
Sea cucumbers with fish			
Beneath net pens (5-mm mesh) of 18 m diameter and 7.5 m depth	*Parastichopus californicus* and salmon fish	1 million salmon fry/ 100 sea cucumber ($\geq$130 cm)	Ahlgren (1998)
Trays (surface area: 1.13 m^2 with 20-mm mesh for large and 5-mm mesh for small size) below fish net pens	Sea cucumber *P. californicus* and sablefish *Anoplopoma fimbria*	Small (<100 g) and large (>100 g), preferably at stocking density of 12 ind/m^2	Hannah et al. (2013)
Under net circular cages of 10 × 10 m and diameter of 1.5 m depth	*Holothuria tubulosa* and finfish sea bass and sea bream	50 sea cucumbers (92.81 ± 2.29) were placed in underwater ranches	Tolon et al. (2017b)
In cages (45 cm diameter and 10 cm height) suspended at 4 m depth	*Holothuria leucospilota* and marine fish species	Sea cucumbers 19.16 ± 1.08 were placed as four sea cucumbers per cage 25 ind/m^2	Yu et al. (2014)
Sea cucumbers with shrimp			
Pens (8 ×8 ×2.5 m) with 0.5-cm mesh size in ponds (about 2 ha) that supplied with plastic tubes as substrates	Sea cucumber *Apostichopus japonicus* and the Chinese white shrimp *Fenneropenaeus chinensis*	Sea cucumbers (5.0 ± 0.2 g) at a density of 10 juveniles/m^2. Hatchery-reared shrimp juvenile of length 1.1 ± 0.5 cm and wet weight 1.2 ± 0.1 g at a density of 3 juveniles/m^2	Zhou et al. (2017)
Co-cultured in cofferdam about 120.2 ha (1881 × 639 m)	Sea cucumber (*A. japonicus*), Jellyfish (*Rhopilema esculenta*) and Shrimp (*F. chinensis*)	Sea cucumber: 9.0 g ± 0.8 g (105 ind./ha); jellyfish: 1.21 g ± 0.5 g (1200 ind/ha); shrimp individuals: 0.098 g ± 0.01 g (59900 ind/ha)	Li et al. (2014)

Continued

Table 6.6 Farming models of sea cucumber with different species.—*cont'd*

IMTA system	Species	Density	References
Sea cucumbers with seaweed			
Cages (1.5 × 1.5 × 0.5 m, L × W × H) using 10-mm polyethylene oyster mesh	The sea cucumber *Holothuria scabra* and the seaweed *Eucheuma denticulatum*	Integration of both species at 200 g/m^2	Namukose et al. (2016)
Under seaweed culture in cages (1.5 × 1.5 × 0.50 m, L × W × H)	Juvenile sea cucumbers, *H. scabra*, and red seaweed *Kappaphycus striatum*	Sea cucumber 124 g/m^2	Beltran-Gutierrez et al. (2016)
Sea cucumbers with other invertebrates			
Beneath cages of (0.90 × 0.90 × 0.23 m)	*Australostichopus mollis* and green-lipped mussel	2.5 and 5 sea cucumber (mean: 125.8 g)	Slater and Carton (2007)
Beneath cages containing abalone inside rearing tanks (55 L)	Abalone *Haliotis discus hannai* and sea cucumber *Stichopus japonicus*	Juvenile (n = 5) (mean: 5.0 g) sea cucumbers were at the bottom of the tank. The juvenile abalone were 0.35 ± 0.12 g (n = 50).	Kang et al. (2003)
Beneath Suspended cages (height: 1 m; mesh: 1 cm) in a pond (100 × 200 × 2 m)	Sea cucumber *A. japonicus* and scallop *Chlamys farreri*	Sea cucumber 5.0 g, 15 ind/m^2. Scallops 2 cm shell height and 4.0 g wet weight were distributed in six layers in each cage; ten in each layer.	Ren et al. (2012)
Trays 56.25 × 56.25 × 21.25 cm (L × W × H) were placed 2.5 m below the suspended oyster	Sea cucumber *P. californicus* and *Crassostrea gigas*	Six Sea cucumber length 8—13 cm were placed in each experimental tray, and oyster 1077 individuals/m^2	Paltzat et al. (2008)
Subsurface united type cage for abalone and sea cucumber grow out with one level. The nominal dimensions were 920 × 1000 × 440 mm^3 and the UC was constructed	*H. discus hannai*, with juvenile sea cucumber, *A. japonicus*	Abalones (n = 500; 2.43 ± 0.45 g) and juvenile sea cucumbers (n = 50) 3.37 ± 1.31 g	Kim et al. (2015)

Table 6.6 Farming models of sea cucumber with different species.—*cont'd*

IMTA system	Species	Density	References
of six sheets of polyethylene plate			
Sea cucumber caged (1.0 × 0.5 × 0.15 m, L × W × H) under oyster by 1.5 m	*A. mollis* and oyster (*C. gigas*)	Juvenile sea cucumbers (36.7 ± 0.9 g, wet weight). Four animals/cage	Zamora et al. (2014)

6.10 Feasibility for sea cucumbers farming

Sea cucumber fisheries have a high potential for socio-economic benefits and worth investing (Eggertsen et al., 2020). The Hong Kong market controls sea cucumber prices. The sea cucumbers *A. japonicus* reach the highest price of US$ 2950/kg dried followed by *H. scabra*, normally US$ 115−640/kg, *Holothuria lessoni* for US$ 240−790/kg and the sea cucumber that belongs to the teat fish group (e.g. *H. fuscogilva*, *Holothuria nobilis* and *Holothuria whitmaei*) can reach to US$130−270/kg dried (Purcell et al., 2012).

Sea cucumbers monoculture has shown promising profitability. In ponds, the carrying capacities are linked to the quality of the substrate, biological productivity and rate of water exchange (Robinson, 2013). The sea cucumber *H. scabra* can achieve survival rates of 80%−87% and growth rates of 1−1.8 g/d, reaching the harvest size within 6−14 months (Robinson, 2013). The growth rate of *H. scabra* likely ranges from 0.1 to 0.9 g/ind. day to 1.0−1.7 g/ind. day (Agudo, 2012; Pitt and Duy, 2004; Purcell and Kirby, 2006; Raison, 2008). In New Caledonia, the growth of the juvenile *H. scabra* in ponds was 20−72 g/month (Agudo, 2012; Purcell, 2004; Purcell et al., 2012). Furthermore, in sea pens, the growths ranged from 1.0 to 1.8 g/d, depending on the carrying capacity of the site, which could reach the market size (minimum 300 g) in 5−8 months (Robinson and Pascal, 2012). Also, the juveniles of the sea cucumber *A. japonicus* take 15−18 months to reach commercial size if the bodyweight of stocked juveniles is less than 1 g (Chen, 2003).

In the co-culture system, many sea cucumber species have proven to grow well under aquaculture farms of bivalves, finfish, macroalgae and crustaceans, achieving a high economic value. Thus, co-culturing sea cucumbers is a promising solution to increase the profitability of the marine farming systems. For instance, the farming of the sea cucumber *Australostichopus mollis* with mussel and oyster can achieve a 2000 kg/ha wet weight. Giving the price of *A. mollis* is US$ 275/kg dry weight, a US$ 44 000 could be obtained per a hectare of *A. mollis* farming in an integrated system (Zamora et al., 2018). Sea cucumber in a rotational cycle with shrimp can yield 2−3 tones/ha y. The rotational monoculture of the sea cucumber with shrimp can achieve high economic returns, ranging from 1700 to 2200 US$/ha crop, with profit margins of 33%−45% (Duy, 2012; Robinson, 2013).

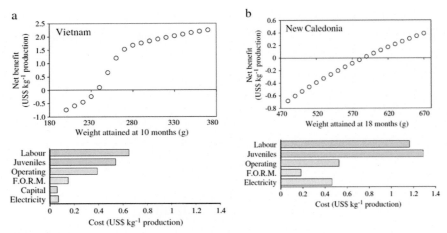

FIGURE 6.12

Plots of modelled net benefit (price received minus production costs) for 8 ha pond farms in (A) Vietnam and (B) New Caledonia. The net benefit is a function of average growth attained over each growing cycle (Vietnam: 10 months; New Caledonia: 18 months). Bar charts show the component costs for the production of 1 kg of sandfish for sizes readily accepted in each country.

Credit: Purcell, S.W., Hair, C.A., Mills, D.J., 2012. Sea cucumber culture, farming and sea ranching in the tropics: progress, problems and opportunities. Aquaculture. https://doi.org/10.1016/j.aquaculture.2012.08.053.

However, the production cost (labour, juvenile price, electricity, pumping and maintenance) should be considered for different culture models under the country condition. For instance, farming the sea cucumber (*H. scabra*) in New Caledonia was unprofitable because of long grow-out period and production costs (Purcell et al., 2012). Prolonging the farming period and increasing the density are possible solutions to increase profitability in that case (Purcell et al., 2012). In China, the sea cucumber *A. japonicus* is packaged separately as 30 sea cucumber weighing 1 kg as a ready-to-eat product. The production cost is around 510 yuan per kg (ca. 75 U.S dollar). Then, the sea cucumbers are packaged and delivered to the market (Personal communication). Furthermore, it is crucial to choose the suitable sea cucumber species that fit the cultured area. For instance, *H. scabra* can grow well in ponds, but not *I. fuscus* or *H. whitmaei*. *H. whitmaei* lost an average of 180 g over 6 months in the same pond of that *H. scabra* gained weight. Also, *H. fuscogilva* does not survive well under co-culturing with milkfish in ponds (Purcell et al., 2012) (Fig. 6.12).

References

Agudo, N., 2006. Sandfish Hatchery Techniques. Secretariat of the Pacific Community, p. 44.

Agudo, N.S., 2012. Pond grow-out trials for sandfish (*Holothuria scabra*) in New Caledonia. Asia—Pacific tropical sea cucumber aquaculture, p. 104.

Ahlgren, M.O., 1998. Consumption and assimilation of salmon net pen fouling debris by the red sea cucumber *Parastichopus californicus*: implications for polyculture. J. World Aquac. Soc. 29 (2), 133—139.

Asha, P.S., Muthiah, P., 2007. Growth of the hatchery-produced juveniles of commercial sea cucumber *Holothuria* (Theelothuria) *spinifera* Theel. Aquacult. Res. 38, 1082—1087. https://doi.org/10.1111/j.1365-2109.2007.01775.x.

Battaglene, S.C., Seymour, J.E., Ramofafia, C., Lane, I., 2002. Spawning induction of three tropical sea cucumbers, *Holothuria scabra, H. fuscogilva* and *Actinopyga mauritiana*. Aquaculture 207, 29—47. https://doi.org/10.1016/S0044-8486(01)00725-6.

Beltran-Gutierrez, M., Ferse, S.C., Kunzmann, A., Stead, S.M., Msuya, F.E., Hoffmeister, T.S., Slater, M.J., 2016. Co-culture of sea cucumber *Holothuria scabra* and red seaweed *Kappaphycus striatum*. Aquacult. Res. 47 (5), 1549—1559.

Cameron, J.L., Fankboner, P.V., 1986. Reproductive biology of the commercial sea cucumber *Parastichopus californicus* (Stimpson) (Echinodermata: Holothuroidea). I. Reproductive periodicity and spawning behavior. Can. J. Zool. 64, 168—175. https://doi.org/10.1139/z86-027.

Cánovas, F., Domínguez-Godino, J.A., González-Wangüemert, M., 2019. Epidemiology of skin ulceration disease in wild sea cucumber *Holothuria arguinensis*, a new aquaculture target species. Dis. Aquat. Org. 135, 77—88. https://doi.org/10.3354/dao03373.

Chao, S.M., Chen, C.P., Alexander, P.S., 1994. Reproduction and growth of *Holothuria atra* (Echinodermata: Holothuroidea) at two contrasting sites in southern Taiwan. Mar. Biol. 119, 565—570. https://doi.org/10.1007/BF00354319.

Chen, J., 2003. Overview of sea cucumber farming and sea ranching practices in China. SPC beche-de-mer Inf. Bull. 18, 18—23.

Chieu, H.D., Turner, L., Smith, M.K., Wang, T., Nocillado, J., Palma, P., Suwansaard, S., Elizur, A., Cummins, S.F., 2019. Aquaculture breeding enhancement: maturation and spawning in sea cucumbers using a recombinant relaxin-like gonad-stimulating peptide. Front. Genet. 10 https://doi.org/10.3389/fgene.2019.00077.

Colombo, L., Pickering, A.D., Belvedere, P., Schreck, C.B., 1989. Stress inducing factors and stress reaction in aquaculture. In: Aquaculture Europe"89 - Business Joins Science, pp. 93—121.

Conand, C., 1981. Sexual cycle of three commercially important holothurian species (Echinodermata) from the Lagoon of New Caledonia. Bull. Mar. Sci. 31, 523—543.

Coulon, P., Jangoux, M., 1993. Feeding rate and sediment reworking by the holothuroid *Holothuria tubulosa* (Echinodermata) in a Mediterranean seagrass bed off Ischia Island. Italy. Mar. Ecol. Prog. Ser. 92, 201—204. https://doi.org/10.3354/meps092201.

Coulon, P., Jangoux, M., Bulteel, P., 1992. Respiratory rate and assessment of secondary production in the holothuroid *Holothuria tubulosa* (Echinodermata) from Mediterranean seagrass beds. Mar. Ecol. 13, 63—68. https://doi.org/10.1111/j.1439-0485.1992.tb00340.x.

Dabbagh, A.-R., Sedaghat, M.R., Rameshi, H., Kamrani, E., 2011. Breeding and larval rearing of the sea cucumber *Holothuria leucospilota* Brandt (*Holothuria vegabunda* Selenka) from the northern Persian Gulf, Iran. SPC Beche-de-mer Inf. Bull. 35—38.

Desurmont, A., 2005. Observations of natural spawning of *Bohadschia vitiensis* and *Holothuria scabra* versicolor. SPC Beche-de-mer Inf. Bull. 98848.

Domínguez-Godino, J.A., González-Wangüemert, M., 2019a. Improving the fitness of *Holothuria arguinensis* larvae through different microalgae diets. Aquacult. Res. 50, 3130–3137. https://doi.org/10.1111/are.14267.

Domínguez-Godino, J.A., González-Wangüemert, M., 2019b. Assessment of *Holothuria arguinensis* feeding rate, growth and absorption efficiency under aquaculture conditions. N. Z. J. Mar. Freshw. Res. 53, 60–76. https://doi.org/10.1080/00288330.2018.1480499.

Domínguez-Godino, J.A., González-Wangüemert, M., 2018. Breeding and larval development of *Holothuria mammata*, a new target species for aquaculture. Aquacult. Res. 49, 1430–1440. https://doi.org/10.1111/are.13597.

Domínguez-Godino, J.A., Slater, M.J., Hannon, C., González-Wangüermert, M., 2015. A new species for sea cucumber ranching and aquaculture: breeding and rearing of *Holothuria arguinensis*. Aquaculture 438, 122–128. https://doi.org/10.1016/j.aquaculture.2015.01.004.

Dong, S., Liang, M., Gao, Q., Wang, F., Dong, Y., Tian, X., 2010. Intra-specific effects of sea cucumber (*Apostichopus japonicus*) with reference to stocking density and body size. Aquacult. Res. 41, 1170–1178. https://doi.org/10.1111/j.1365-2109.2009.02404.x.

Duy, N.D.Q., 2010. Seed production of sandfish (*Holothuria scabra*) in Vietnam. Aquaculture Department, Southeast Asian Fisheries Development Center.

Duy, N.D., 2012. Large-scale sandfish production from pond culture in Vietnam. Asia–Pac. Trop. Sea Cucumber Aquacult. ACIAR Proc. 136, 34–39.

Eggertsen, M., Eriksson, H., Slater, M.J., Raymond, C., de la Torre-Castro, M., 2020. Economic value of small-scale sea cucumber fisheries under two contrasting management regimes. Ecol. Soc. 25, 1–10. https://doi.org/10.5751/ES-11436-250220.

Ghobadyan, F., Morovvati, H., Ghazvineh, L., Tavassolpour, E., 2012. An investigation of the macroscopic and microscopic characteristics\r of gonadal tubules in the sea cucumber *Holothuria leucospilota* (Persian Gulf, Iran). SPC Beche Mer Inf. Bull. 6–14.

Günay, D., Emiroğlu, D., Suzer, C., 2020. Seasonal variations of digestive enzymes in sea cucumbers (*Holothuria tubulosa*, G. 1788) under culture conditions. J. Exp. Zool. Part A Ecol. Integr. Physiol. 333, 144–150. https://doi.org/10.1002/jez.2336.

Günay, D., Emiroğlu, D., Tolon, T., Özden, O., Saygi, H., 2015. Farklı sıcaklıklarda deniz hıyarı (*Holothuria tubulosa*, Gmelin, 1788) genç bireylerinin büyüme ve yaşama oranı. Turk. J. Fish. Aquat. Sci. 15, 533–541. https://doi.org/10.4194/1303-2712-v15_2_41.

Günay, D., Tolon, M.T., Emiroğlu, D., 2018. Effects of various stocking densities on hatching and survival rates of sea cucumber *Holothuria tubulosa* eggs (Gmelin, 1788). Ege J. Fish. Aquat. Sci. 35, 381–386. https://doi.org/10.12714/egejfas.2018.35.4.03.

Hamel, J.F., Mercier, A., 1996. Early development, settlement, growth, and spatial distribution of the sea cucumber *Cucumaria frondosa* (Echinodermata: Holothuroidea). Can. J. Fish. Aquat. Sci. 53, 253–271. https://doi.org/10.1139/f95-186.

Hannah, L., Pearce, C.M., Cross, S.F., 2013. Growth and survival of California sea cucumbers (*Parastichopus californicus*) cultivated with sablefish (*Anoplopoma fimbria*) at an integrated multi-trophic aquaculture site. Aquaculture 406, 34–42.

Hu, C., Li, H., Xia, J., Zhang, L., Luo, P., Fan, S., Peng, P., Yang, H., Wen, J., 2013. Spawning, larval development and juvenile growth of the sea cucumber *Stichopus horrens*. Aquaculture 404–405, 47–54. https://doi.org/10.1016/j.aquaculture.2013.04.007.

Huang, W., Huo, D., Yu, Z., Ren, C., Jiang, X., Luo, P., Chen, T., Hu, C., 2018. Spawning, larval development and juvenile growth of the tropical sea cucumber *Holothuria leucospilota*. Aquaculture 488, 22–29. https://doi.org/10.1016/j.aquaculture.2018.01.013.

Kang, K.H., Kwon, J.Y., Kim, Y.M., 2003. A beneficial coculture: charm abalone *Haliotis discus* hannai and sea cucumber *Stichopus japonicus*. Aquaculture 216, 87–93. https://doi.org/10.1016/S0044-8486(02)00203-X.

Kim, T., Yoon, H.S., Shin, S., Oh, M.H., Kwon, I., Lee, J., Choi, S.D., Jeong, K.S., 2015. Physical and biological evaluation of co-culture cage systems for grow-out of juvenile abalone, *Haliotis discus hannai*, with juvenile sea cucumber, *Apostichopus japonicus* (Selenka), with CFD analysis and indoor seawater tanks. Aquaculture 447, 86–101.

Lavitra, T., Rasolofonirina, R., Eeckhaut, I., 2010. The effect of sediment quality and stocking density on survival and growth of the sea cucumber *Holothuria scabra* reared in nursery ponds and sea pens. West. Indian Ocean J. Mar. Sci. 9, 153–164.

Lawrence, J., 1982. Echinoderms. CRC Press.

Laxminarayana, A., 2005. Induced spawning and larval rearing of the sea cucumbers, *Bohadschia marmorata* and *Holothuria atra* in Mauritius. Fish. Res. 48–52.

Li, H., Luo, P., Yu, Z., Hu, C., 2013. Preliminary feasibility study of cage culture of tropical sea cucumber (*Stichopus horrens*) in the wide. South China Fish. Sci. 9, 1–7. https://doi.org/10.3969/j.issn.2095-0780.2013.06.001.

Li, J., Dong, S., Gao, Q., Zhu, C., 2014. Nitrogen and phosphorus budget of a polyculture system of sea cucumber (*Apostichopus japonicus*), jellyfish (*Rhopilema esculenta*) and shrimp (Fenneropenaeus chinensis). J. Ocean Univ. China 13, 503–508. https://doi.org/10.1007/s11802-014-2181-9.

MacDonald, C.L.E., Stead, S.M., Slater, M.J., 2013. Consumption and remediation of European Seabass (*Dicentrarchus labrax*) waste by the sea cucumber *Holothuria forskali*. Aquacult. Int. 21, 1279–1290. https://doi.org/10.1007/s10499-013-9629-6.

Marquet, N., Conand, C., Power, D.M., Canário, A.V., González-Wangüemert, M., 2017. Sea cucumbers, *Holothuria arguinensis* and *H. mammata*, from the southern Iberian Peninsula: variation in reproductive activity between populations from different habitats. Fish. Res. 191, 120–130.

McEuen, F.S., 1988. Spawning behaviors of northeast Pacific sea cucumbers (Holothuroidea: Echinodermata). Mar. Biol. 98, 565–585. https://doi.org/10.1007/BF00391548.

Mu, J., Song, J., 2005. Technology of the raft caged *Apostichopus japonicus* aquaculture in epicontinental seas. Sci. Fish Farming 3, 39.

Namukose, M., Msuya, F.E., Ferse, S.C.A., Slater, M.J., Kunzmann, A., 2016. Growth performance of the sea cucumber *Holothuria scabra* and the seaweed *Eucheuma denticulatum*: integrated mariculture and effects on sediment organic characteristics. Aquacult. Environ. Interact. 8, 179–189. https://doi.org/10.3354/aei00172.

Ocaña, A., Sánchez, T.L., 2005. Spawning of *Holothuria tubulosa* (Holothurioidea, Echinodermata) in the Alboran Sea (Mediterranean Sea). Zool. Baetica 16, 147–150.

Olaya-Restrepo, J., Erzini, K., González-Wangüemert, M., 2018. Estimation of growth parameters for the exploited sea cucumber *holothuria arguinensis* from south Portugal. Fish. Bull. 116, 1–8. https://doi.org/10.7755/FB.116.1.1.

Palomar-Abesamis, N., Juinio-Meñez, M.A., Slater, M.J., 2018a. Macrophyte detritus as nursery diets for juvenile sea cucumber *Stichopus* cf. *horrens*. Aquacult. Res. 49, 3614–3623. https://doi.org/10.1111/are.13829.

Palomar-Abesamis, N., Juinio-Meñez, M.A., Slater, M.J., 2018b. Effects of light and microhabitat on activity pattern and behaviour of wild and hatchery-reared juveniles of *Stichopus* cf. *horrens*. J. Mar. Biol. Assoc. U. K. 98, 1703–1713. https://doi.org/10.1017/S0025315417000972.

Paltzat, D.L., Pearce, C.M., Barnes, P.A., McKinley, R.S., 2008. Growth and production of California sea cucumbers (*Parastichopus californicus* Stimpson) co-cultured with suspended Pacific oysters (*Crassostrea gigas* Thunberg). Aquaculture 275 (1–4), 124–137.

Peng, P., Hu, C., Yu, Z., Luo, P., 2012. Preliminary study on purification of sediments in cage culture by *Holothuria leucospilota* and *Sargassum hemiphyllum*. Mar. Environ. Sci. 31, 316–322.

Pitt, R., Duy, N., Duy, T., Long, H., 2004. Sandfish (Holothuria scabra) with shrimp (*Penaeus monodon*) co-culture tank trials. SPC Beche-de-mer Inf. Bull. 12–22.

Pitt, R., Duy, N.D.Q., 2004. Breeding and rearing of the sea cucumber *Holothuria scabra* in Vietnam. In: Advances in Sea Cucumber Aquaculture and Management, pp. 333–346.

Purcell, 2004. Criteria for release strategies and evaluating the restocking of sea cucumbers. Advances in sea cucumber aquaculture and management. FAO.

Purcell, S.W., Hair, C.A., Mills, D.J., 2012. Sea cucumber culture, farming and sea ranching in the tropics: progress, problems and opportunities. Aquaculture. https://doi.org/10.1016/j.aquaculture.2012.08.053.

Purcell, S.W., Kirby, D.S., 2006. Restocking the sea cucumber *Holothuria scabra*: Sizing no-take zones through individual-based movement modelling. Fish. Res. 80 (1), 53–61.

Purcell, S.W., Patrois, J., Fraisse, N., 2006. Experimental evaluation of co-culture of juvenile sea cucumbers, *Holothuria scabra* (Jaeger), with juvenile blue shrimp, *Litopenaeus stylirostris* (Stimpson). Aquacult. Res. 37, 515–522. https://doi.org/10.1111/j.1365-2109.2006.01458.x.

Purcell, S.W., Simutoga, M., 2008. Spatio-temporal and size-dependent variation in the success of releasing cultured sea cucumbers in the wild. Rev. Fish. Sci. 16, 204–214. https://doi.org/10.1080/10641260701686895.

Purwati, P., 2003. Natural Spawning Observations of *Pearsonothuria graeffei* coastfish.spc.int 98848.

Raison, C.M., 2008. Advances in sea cucumber aquaculture and prospects for commercial culture of Holothuria scabra. CAB Rev.: Perspec. Agri. Veter. Sci. Nut. Nat. Res. 3 (082), 1–15.

Rakaj, A., Fianchini, A., Boncagni, P., Lovatelli, A., Scardi, M., Cataudella, S., 2018. Spawning and rearing of *Holothuria tubulosa*: a new candidate for aquaculture in the Mediterranean region. Aquacult. Res. 49, 557–568. https://doi.org/10.1111/are.13487.

Rakaj, A., Fianchini, A., Boncagni, P., Scardi, M., Cataudella, S., 2019. Artificial reproduction of *Holothuria polii*: a new candidate for aquaculture. Aquaculture 498, 444–453. https://doi.org/10.1016/j.aquaculture.2018.08.060.

Ramofafia, C., Battaglene, S.C., Bell, J.D., Byrne, M., 2000. Reproductive biology of the commercial sea cucumber *Holothuria fuscogilva* in the Solomon Islands. Mar. Biol. 136 (6), 1045–1056.

Ramofafia, C., Byrne, M., Battaglene, S., 2001. Reproductive biology of the intertidal sea cucumber *Actinopyga mauritiana* in the Solomon Islands. J. Mar. Biol. Assoc. U. K. 81, 523–531. https://doi.org/10.1017/s0025315401004179.

Ramofafia, C., Gervis, M., Bell, J., 1995. Spawning and early larval rearing of *Holothuria atra*. Beche-de-mer Inf. Bull. 7.

Ren, Y., Dong, S., Qin, C., Wang, F., Tian, X., Gao, Q., 2012. Ecological effects of co-culturing sea cucumber *Apostichopus japonicus* (Selenka) with scallop *Chlamys farreri* in earthen ponds. Chin. J. Oceanol. Limnol. 30, 71–79. https://doi.org/10.1007/s00343-012-1038-6.

Robinson, G., 2013. A bright future for sandfish aquaculture. World Aquacult. Soc. Mag. 18–24.

Robinson, G., Pascal, B., 2012. Sea cucumber farming experiences in. Asia–Pacific trop. Sea cucumber aquac. ACIAR Proc. 142–155.

Santos, R., Dias, S., Tecelão, C., Pedrosa, R., Pombo, A., 2017. Reproductive biological characteristics and fatty acid profile of *Holothuria mammata* (Grube, 1840). SPC Beche-de-mer Inf. Bull. 37, 57–64.

Sewell, M.A., Levitan, D.R., 1992. Fertilization success during a natural spawning of the dendrochirote sea cucumber *Cucumaria miniata*. Bull. Mar. Sci. 51, 161–166.

Slater, M.J., Carton, A.G., 2007. Survivorship and growth of the sea cucumber *Australostichopus* (Stichopus) *mollis* (Hutton 1872) in polyculture trials with green-lipped mussel farms. Aquaculture 272, 389–398. https://doi.org/10.1016/j.aquaculture.2007.07.230.

Slimane-Tamacha, F., Soualili, D., Mezali, K., 2019. Reproductive biology of *Holothuria* (Roweothuria) *poli* (Holothuroidea: Echinodermata) from Oran Bay, Algeria. SPC Beche-de-mer Inf. Bull. 47–53.

Tolon, M.T., Emiroglu, D., Gunay, D., Ozgul, A., 2017b. Sea cucumber (*Holothuria tubulosa* Gmelin, 1790) culture under marine fish net cages for potential use in integrated multi-trophic aquaculture (IMTA). Indian J. Geo Mar. Sci. 46, 749–756.

Tolon, M.T., Engin, S., 2019. Gonadal development of the holothurian *Holothuria polii* (Delle chiaje, 1823) in spawning period at the Aegean sea (Mediterranean sea). Ege J. Fish. Aquat. Sci. 36, 379–385. https://doi.org/10.12714/egejfas.36.4.09.

Tolon, T., 2017. Effect of salinity on growth and survival of the juvenile sea cucumbers *Holothuria tubulosa* (Gmelin, 1788) and *Holothuria poli* (Delle Chiaje, 1923). Fresenius Environ. Bull. 26 (6), 3930–3935.

Tolon, T., Emiroğlu, D., Günay, D., Hancı, B., 2017a. Effect of stocking density on growth performance of juvenile sea cucumber *Holothuria tubulosa* (Gmelin, 1788). Aquacult. Res. 48, 4124–4131. https://doi.org/10.1111/are.13232.

Tolon, T., Emiroğlu, D., Günay, D., Saygı, H., 2015. Sediment tane boyutunun deniz hıyarı (*Holothuria tubulosa*) genç bireylerinin büyüme performansı üzerine etkileri. Turk. J. Fish. Aquat. Sci. 15, 555–559. https://doi.org/10.4194/1303-2712-v15_2_43.

Toral-Granda, V., 2008. Galapagos Islands: a hotspot of sea cucumber fisheries in Latin America and the Caribbean. Sea cucumbers: A global review of fisheries and trade, pp. 231–253.

Wang, J.Q., Cheng, X., Gao, Z.Y., Chi, W., Wang, N.B., 2008. Primary results of polyculture of juvenile sea cucumber (*Apostichopus japonicus* Selenka) with juvenile sea urchin (*Strongylocentrotus intermedius*)and manila clam (*Ruditapes philippinarum*). J. Fish. China/Shuichan Xuebao 32, 740–748.

Wang, J.Q., Cheng, X., Yang, Y., Wang, N.B., 2007. Polyculture of juvenile sea urchin (Strongylocentrotus intermedius) with juvenile sea cucumber (Apostichopus japonicus Selenka) at various stocking densities. J. Dalian Fish. Univ. 2, 102–108.

Yokoyama, H., 2013. Growth and food source of the sea cucumber *Apostichopus japonicus* cultured below fish cages—potential for integrated multi-trophic aquaculture. Aquaculture 372, 28–38.

Yu, Z.-he, Hu, C., Qi, Z., Jiang, H., Ren, C., Luo, P., 2012a. Co-culture of sea cucumber *Holothuria leucospilota* with the Pacific white shrimp *Litopenaeus vannamei*. J. Fish. China 36, 1081. https://doi.org/10.3724/sp.j.1231.2012.27864.

Yu, Z., Hu, C., Zhou, Y., Li, H., Peng, P., 2012b. Survival and growth of the sea cucumber *Holothuria leucospilota* Brandt: a comparison between suspended and bottom cultures in a subtropical fish farm during summer. Aquacult. Res. 44, 114–124. https://doi.org/10.1111/j.1365-2109.2011.03016.x.

Yu, Z., Qi, Z., Hu, C., Liu, W., Huang, H., 2013. Effects of salinity on ingestion, oxygen consumption and ammonium excretion rates of the sea cucumber *Holothuria leucospilota*. Aquacult. Res. 44, 1760–1767. https://doi.org/10.1111/j.1365-2109.2012.03182.x.

Yu, Z., Zhou, Y., Yang, H., Hu, C., 2014. Survival, growth, food availability and assimilation efficiency of the sea cucumber *Apostichopus japonicus* bottom-cultured under a fish farm in southern China. Aquaculture 426, 238–248.

Yuan, X.T., Yang, H.S., Zhou, Y., Mao, Y.Z., Xu, Q., Wang, L.L., 2008. Bioremediation potential of *Apostichopus japonicus* (Selenka) in coastal bivalve suspension aquaculture system. Chin. J. Appl. Ecol. 19, 866–872.

Zamora, L.N., Dollimore, J., Jeffs, A.G., 2014. Feasibility of co-culture of the Australasian sea cucumber (*Australostichopus mollis*) with the Pacific oyster (*Crassostrea gigas*) in northern New Zealand. N. Z. J. Mar. Freshwater Res. 48 (3), 394–404.

Zamora, L.N., Yuan, X., Carton, A.G., Slater, M.J., 2018. Role of deposit-feeding sea cucumbers in integrated multi-trophic aquaculture: progress, problems, potential and future challenges. In: Reviews in Aquaculture. Wiley-Blackwell. https://doi.org/10.1111/raq.12147.

Zhou, S., Ren, Y., Pearce, C.M., Dong, S., Tian, X., Gao, Q., Wang, F., 2017. Ecological effects of co-culturing the sea cucumber *Apostichopus japonicus* with the Chinese white shrimp Fenneropenaeus chinensis in an earthen pond. Chin. J. Oceanol. Limnol. 35, 122–131. https://doi.org/10.1007/s00343-016-5184-0.

Sea cucumbers processing and cooking

CHAPTER

7

7.1 Processing sea cucumbers

A variety of processing methods were developed to process sea cucumbers into bêche-de-mer since sea cucumbers are autolysed after being harvested (Fig. 7.1). Prolonged exposure of fresh sea cucumber to sunlight can cause biochemical reactions and radical activation (Ravinesh Ram, 2014). Once processed, sea cucumbers can be stored for a long time (Figs. 7.2 and 7.3).

7.1.1 Drying

Food drying has been around since ancient times for food preservation in which water is removed from the food. Drying the food helps to preserve flavour and colour, inhibit the growth of microorganisms and reduce storage transportation costs. Natural energies, such as wind and sunlight, have been utilised from ancient times to

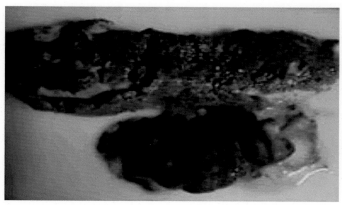

FIGURE 7.1

Flesh deterioration of the sea cucumber *Stichopus hermanni* after harvesting.
Credit: Ravinesh Ram, P.C., 2014. Effects of harvest and post-harvest processing methods on quality of beche-de-mer in Fiji Islands. J. Mar. Sci. Res. Dev. 04, 1–6. https://doi.org/10.4172/2155-9910.1000153.

Sea Cucumbers. https://doi.org/10.1016/B978-0-12-824377-0.00011-6

FIGURE 7.2

Dried sea cucumbers for sale.

Credit: Bare Dreamer. Licensed under CC BY 2.0 via Wikimedia Commons.

reduce the moisture content of the food. The traditional method of dehydrating sea cucumbers involves gutting, boiling and smoking and sun drying. The sea cucumbers are firstly boiled (2−4 min) then the internal organs are removed (gutted) and then boiled again until hard and stiff (Chong et al., 2015). Also, Subaldo (2011) reported that the internal organs of sea cucumbers are removed, and then, they are boiled in water (ca. 30 min.) (Subaldo, 2011). However, this method has shortcomings, including time and reliable quality. Therefore, scientists sought for new drying technologies to introduce fresher, safer and higher quality food, with

FIGURE 7.3

Dried sea cucumbers ready for exportation.

Credit: Ahmed, M., 2009. Morphological, Ecological and Molecular Examination of the Sea Cucumber Species along the Red Seacoast of Egypt and Gulf of Aqaba. The University of Hull.

more extended life for marketing (Hayashi, 1989). Several drying technologies were developed to yield high-quality dehydrated sea cucumbers, such as microwave drying and freeze-drying.

7.1.1.1 Solar drying: an ancient method worth development

Drying using the sun and under the presence of wind has been practised since ancient times to preserve food. This method does not require a machine and/or energy costs, so it is cheap. Therefore, fishers can do it in a suitable climate, and the resulting product is transported to the market (Lavitra et al., 2008; Rasolofonirina et al., 2004; Roveena Vandana Chand, 2014; Schoppe, 2000). Sun-dried sea cucumber products have acceptable sensory aspects of taste and appearance. However, sea cucumber products using this method are more likely to be invaded by pests, birds and rodents (Chong et al., 2015). Vaipulu (2009) developed a solar dryer using solar panels for drying sea cucumber, which has its advantages over the traditional open sun drying method, such as reducing the loss due to damage caused by insects, birds, rodents and adverse climatic conditions. The drying period using the conceptual solar dryer is 1−2 d, whereas it takes 4−14 d in the traditional open sun drying method (Vaipulu, 2009).

7.1.1.2 Oven drying

One of the drying methods used to dry sea cucumber is oven drying. Bilgin and Öztürk Tanrikulu (2018) compared the nutritional properties of fresh, boiled, ambient-dried ($23 \pm 2°C$) and oven-dried ($45 \pm 1°C$) sea cucumber *Holothuria tubulosa* in terms of proximate composition and fatty acid profiles. The authors concluded that oven drying is better than the ambient drying in terms of preservation, whereas drying at the ambient temperature is better in terms of nutrient value (Bilgin and Öztürk Tanrikulu, 2018). In another study using the Mediterranean sea cucumber *Holothuria forskali*, Telahigue et al. (2014) used a controlled oven dryer system, which resulted in the improvement of the nutritional quality of the sea cucumber under the temperature of 60°C and humidity of 20% (Telahigue et al., 2014).

7.1.1.3 Hot air drying

Hot air drying has been employed as a mature technology in which the heat is transferred to the product and evaporates the water (Antal, 2015). Li et al. (2018) reported that hot air drying at 60°C could meet the demand of high nutritional quality of dried sea cucumbers *Apostichopus japonicus* at low cost, compared with sun drying (18°C−25°C), vacuum freeze-drying (−85 to −90°C, and 5−10 Pa vacuum degree, 0.5 cm thickness material and −25°C material temperature), double-distilled water boiling followed by sun drying and 3.5% sodium chloride solution boiling followed by sun drying (Li et al., 2018). However, drying with only hot airflow takes a long time and might cause deterioration of the final product. Duan et al. (2007) used hot airflow through uniformly spread sea cucumber on a mesh, with a velocity of 1.5 m/s and 20% relative humidity and temperature of 60°C. Sea cucumbers were dehydrated until they reached the final moisture content [7% wet basis (w.b.)]. The researchers reported that air-dried sea cucumbers had the worst quality. Also, this method had a long drying time compared with other drying methods of vacuum-microwave drying, freeze-drying and microwave drying (Duan et al., 2007). Furthermore, Moon et al. (2014) dried the sea cucumber *A. japonicus* using far-infrared radiation drying (FIRD) and hot air drying (Moon et al., 2014). The authors concluded that the far-infrared radiation dryer allows food drying in which the dried product absorbs the electromagnetic wave energy. The quality of dried sea cucumber using FIRD was better than that of air drying because the radiation heating mechanism minimised the surface hardening of the sea cucumber (Moon et al., 2014). Therefore, FIRD might be a superior technology to replace solar drying as well as the air drying to produce the dried sea cucumber (Moon et al., 2014).

7.1.1.4 Microwave drying

Microwave drying is used in food drying in which the microwaves penetrate the material and converted to heat that allows the removal of the moisture (Drouzas and Schubert, 1996). This technology has many features, such as rapid heating, high efficiency, good controllability and sanitation (Zhang et al., 2010). Öztürk and Gündüz (2018) placed the sea cucumber *H. tubulosa* in 200 W microwaves for 240 s, achieving a moisture level of 6%. Comparing with hot air and freeze-drying, microwave drying resulted in a high-quality dried product, preserving the amino acids and fatty acids content with short drying time (Öztürk and Gündüz, 2018).

Microwave drying has been combined with another method to increase the efficiency of drying: microwave-assisted vacuum drying or microwave-vacuum drying. This technology not only has the advantages of microwave heating but also lowers the boiling point of water caused by the vacuum environment, thus improving the energy efficiency and decreasing the formation of burned spots in the surface of the final products (Zhang et al., 2010). Zhang et al. (2012) reported that microwave power density of 2 W/g, degree of vacuum of 0.090 MPa and initial cooking water salinity of 80 g/L were optimal conditions for drying the sea cucumber *A. japonicus*. The dried method results in a high rehydration rate (266.32%) and a low shrinkage rate (32.20%), with 110 min drying time.

7.1.1.5 Freeze-drying

Freeze-drying accomplishes water removal from products by sublimation at low pressure and temperature. Compared with other dehydration methods, the freeze-drying has many features, such as negligible loss of nutrients, high rehydration capability and retaining original structure and colour (Qian et al., 2012). However, because of the long drying time and high energy consumption, researchers sought better alternatives (Duan et al., 2007, 2010b). Therefore, freeze-drying is assisted with microwave or vacuum-microwave drying (Duan et al., 2010).

Microwave freeze-drying (MFD) is a freeze-drying operation that is assisted by a microwave field, which is used to supply the heat of sublimation needed. Comparing with the freeze-drying method, MFD offers less drying time, energy consumption and improved the product quality (Duan et al., 2010; Zhang et al., 2010). Duan et al. (2010b) reported that MFD reduces the drying time by about half of the conventional freezing process and provides a similar product quality. The sea cucumber (*A. japonicus*) samples were firstly frozen at $-20°C$ for at least 8 h then dried to 7% w.b. moisture content in the MFD. The optimal conditions for a high-quality product from MFD were an absolute pressure of 50 Pa, a microwave power level of 2 W/g and a microwave cold trap temperature of $-40°C$ (Duan et al., 2010b).

Furthermore, freeze and microwave-vacuum combination can produce high-quality dried sea cucumber with a high rehydration rate and low shrinkage rate, short production time and low energy consumption (Duan et al., 2007; Qian et al., 2012). Qian et al. (2012) reported that freeze and microwave-vacuum combination drying technique of the fresh sea cucumbers *A. japonicus* resulted in a high-quality product in terms of nutrient retaining (Qian et al., 2012). The optimum conditions were 8 h freeze-drying conversion point, 1.7 W/g initial microwave power density and 12 s on and 18 s off microwave intermittent ratio (Qian et al., 2012).

Drying pretreatments were developed to improve the quality and reduce the drying time of freeze-assisted vacuum or microwave drying. Bai and Luan (2018) reported that the vacuum freeze-drying experiments for sea cucumbers pretreated with a high-pulsed electric field (HPEF) could significantly reduce the drying time, save drying energy and improve the rehydration rate of sea cucumber, without significant effect on the product quality. The optimum parameter of HPEF pretreatment was 22.5 kV for 70 Hz, under which the drying time and the energy consumption could be reduced (Bai and Luan, 2018). Also, MFD combined with nanoscale silver coating treatment reduced microorganisms' load, without a significant effect on the product quality (Duan et al., 2008). Moreover, vacuum impregnation with nanoscale calcium carbonate combined with microwave freeze-drying was found to be an efficient drying method for sea cucumber. Compared with MFD method without any treatments, this drying method could reduce the drying time by up to 2 hours (Duan et al., 2010a).

7.1.1.6 Electrohydrodynamic drying

Electrohydrodynamic (EHD) drying is a novel non-thermal technique. It is applied by inducing electric wind (i.e. corona wind), which is generated by gaseous ions under the influence of high voltage electric field. This process is useful for drying heat-sensitive materials because it involves no direct heat (Defraeye and Martynenko, 2018). Bai et al. (2013) compared EHD drying of sea cucumbers at $18°C$ with two drying methods of sea cucumbers: ambient air drying at $18°C$ and oven drying at $80°C$. EHD drying was better than the other two methods in terms of high quality and energy consumption. EHD drying consumes less energy and yields superior quality in terms of physicochemical characters, such as shrinkage, rehydration rate, protein and acid mucopolysaccharide content, texture, colour, appearance and flavour (Bai et al., 2013).

A combination of EHD and vacuum freeze-drying was developed as an improved method for drying sea cucumbers. The combined process resulted in less drying time and has lower energy consumption. Also, the product displays low shrinkage, high rehydration rate and high protein content, along with high quality (Bai et al., 2012).

7.1.1.7 Dry-salting sea cucumbers

A dry-salting method of sea cucumbers includes gutting, cooking, salting, drying and packing (Hernández et al., 2017; Moon and Yoon, 2015).

To process the sea cucumber as salted and dried, the sea cucumber is gutted [by removing the viscera through making an opening (preferably small) on the posterior ventral surface]. Then, sea cucumbers are boiled in the water for 40 min, and then, filter papers are used to remove the water on the surface of the sea cucumber. After that, the sea cucumber is salted 1:1 (sea cucumbers: salt weight) for 48 h. Then, sea cucumbers are dried. Hernández et al. (2017) dried the salted sea cucumber (*Isostichopus* sp. *aff badionotus*) at 60°C, which consumed the shortest time (67.5 h) compared with drying at 50 °C or sun drying. However, drying the sea cucumber at 50°C yielded the highest protein content (Hernández et al., 2017).

7.1.2 Instant sea cucumbers

The development of packaged instant sea cucumbers provided many features, such as high-quality products in terms of nutritional values and eases the marketing process as well as attracts consumers (Liu et al., 2016; Meng et al., 2017).

To process the sea cucumber as ready-to-eat, the sea cucumbers are first gutted then cleaned with distilled water. After that, the sea cucumbers are placed in 85–95°C water for 2 min. Then, they are placed in a refrigerator at −20°C. After that, the sea cucumbers are boiled in a sauce (i.e. flavoured liquid; 1:3, sea cucumber: flavoured liquid) for 17 min, and then, sea cucumbers are soaked in distilled water. Subsequently, the surface moisture is removed from the sea cucumber, packed with a packing machine and sterilised (Liu et al., 2016; Meng et al., 2017).

Two ways of producing ready-to-eat sea cucumbers were compared (Li et al., 2019): traditional ready-to-eat sea cucumbers and vacuum cooking (Fig. 7.4). The

FIGURE 7.4

Ready-to-eat sea cucumbers from the Chinese online shops.

Credit: Mohamed Mohsen.

researchers reported the first method as follows: the body wall of the sea cucumbers is boiled in water for about 20 min, followed by salting for 12 h (3:1 w/w). After that, the salted body wall is boiled in freshwater for about 90 min and then soaked for 20 h in 4°C distilled water. Finally, the samples are boiled for 20 min in purified water, followed by soaking for 40 h. The second method was reported as follows: after boiling the body walls for 15 min in freshwater, the samples are subjected to vacuum cooking machine for 3 h, with a heating temperature of jacketed kettle 95°C and vacuum pressure of vacuum pan of −0.04 MPa. Finally, sea cucumbers are cleaned with purified water to produce the ready-to-eat sea cucumbers (Li et al., 2019). The authors from this comparison concluded that vacuum cooking without soaking might be a promising alternative for producing ready-to-eat sea cucumber products with the high nutritional quality compared with the traditional ready-to-eat method (Fig. 7.5) (Li et al., 2019).

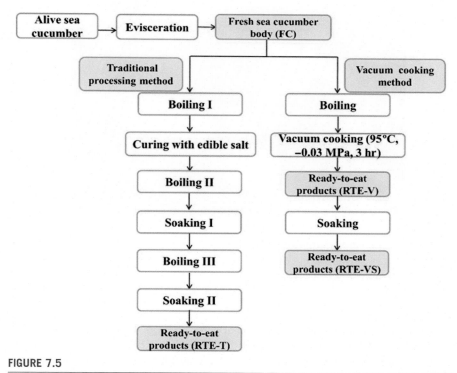

FIGURE 7.5

The processing procedures of ready-to-eat sea cucumbers.

Credit: Li, M., Qi, Y., Mu, L., Li, Z., Zhao, Q., Sun, J., Jiang, Q., 2019. Effects of processing method on chemical compositions and nutritional quality of ready-to-eat sea cucumber (Apostichopus japonicus)*. Food Sci. Nutr. 7, 755–763. https://doi.org/10.1002/fsn3.921.*

7.1.3 Canned sea cucumbers

Canning for food preservation is a useful technology in which the food is processed and sealed. Canned sea cucumbers are firstly gutted, soaked in water for 12—24 h, cooked under high pressure of 0.05—0.15 MPa for 15—35 min and then, sea cucumbers are installed in −4—4°C water for 20—24 h (Jianrong et al., 2007). Pan et al. (2014) produced the canned sea cucumbers as follows: sea cucumbers are gutted then boiled in water at 90°C for 30 s. After that, sea cucumbers are soaked in 10 volumes of sodium phytate and sodium tripolyphosphate, preferably for 3 h to inhibit the shrinkage of sea cucumbers. Then, sea cucumbers are canned, and sodium phytate and sodium tripolyphosphate are added. After that, the canned sea cucumbers are sealed and sterilised at 121°C for 20 min and then cooled and stored (Fig. 7.6) (Pan et al., 2014).

FIGURE 7.6

Canned sea cucumber from Chinese online shops.

Credit: Mohamed Mohsen.

7.2 Cooking

The dried sea cucumbers are rehydrated first before cooking, optimally in pure iced freshwater until softening (Fig. 7.7). Rehydration can take several days (1—4 d), and the water is changed daily. If sea cucumbers are still hard, they can be boiled for 30 min. They are then left in pure freshwater from 1 to 3 d.

Sea cucumbers are cleaned through gentle scraping on the surface. Then, several methods can be followed to cook the sea cucumber. Sea cucumber can be cut into pieces and then fried in oil with mushrooms or with other favourite vegetables;

FIGURE 7.7

Sea cucumbers in a restaurant.

Credit: Mohamed Mohsen.

then, the favourite sauce can be added. Also, the sea cucumber can be boiled in a cooked gruel of rice with the favourite sauce. Furthermore, the sea cucumber can be steamed and introduced with the favourite sauce. The sea cucumber also can be cut into slices and introduced with sauce and vegetables (see below).

7.2.1 Braised sea cucumbers

Sea cucumbers are cut into pieces and stir-fried for about 3 min for later use. Shrimp are dried until golden in colour, and then, mushrooms are added (other vegetables can be added as well) to fry for a few minutes. After that, pieces of sea cucumbers together with the favourite sauce (e.g. soy sauce, oyster sauce, dark caramel sauce, etc.) are added. Then, hot water is added and boiled until the mixture is thick (about 5 min). After that, the mixture can be fried until simmering (Fig. 7.8) (Liew, 2015).

FIGURE 7.8

Braised sea cucumber dish.

Credit: Mohamed Mohsen.

7.2.2 Sea cucumber salad

After softening the sea cucumber, they can be cut into slices (0.25 inch) and wakame seaweeds are soaked in warm water for 3—5 min. Then, they are mixed with carrot, sesame seeds, some lemon juice, dried onion, some ginger and some salt (Fig. 7.9) (Jenny and Heather, 2016).

7.2.3 Sea cucumber with mushrooms

Sea cucumbers are rehydrated in water, which might spend up to 4 d. After the first day, sea cucumbers can be cooked in a saucepan of simmering water for 1 h and then soaked again overnight. Cooking and soaking can be repeated until softened. Once softened, the sea cucumbers are scrapped gently from inside and cut into chunks for later use. Then, mushrooms are prepared by placing them in a saucepan with water and oil for 1 h until simmering. After that, chicken and egg white are combined with one tablespoon of the cornflour and stir-fried. Finally, sea cucumbers, mushrooms, chicken and the favourite sauce (e.g. soy sauce and oyster sauce) are combined and cooked for 2 min (Fig. 7.10) (Veneracion, 2018).

7.2.4 Deep-fried sea cucumber

After rehydration, sea cucumbers are rinsed with some salt, pepper and egg for 15 min. Then, sea cucumbers are placed in crispy butter (frying powder, clustered powder, oil and ice water) until golden brown (Fig. 7.11).

FIGURE 7.9

Sea cucumber with vegetables and mushrooms.

Credit: By avlxyz. Licensed under CC BY-SA 2.0 via Wikimedia Commons.

FIGURE 7.10

Chinese Indonesian sea cucumber with mushroom.

Credit: Gunawan Kartapranata. Licensed under CC BY-SA 4.0 via Wikimedia commons.

FIGURE 7.11

Deep-fried sea cucumbers.

Credit: By Kent Wang. Licensed under CC BY- SA 2.0 via Wikimedia Commons.

7.2.5 Braised sea cucumbers with vegetables

After rehydration of the sea cucumber, they are boiled in water for 2−5 min and then drained and set aside. Next, onion and ginger are stir-fried until they start to release their fragrance. Then, sea cucumbers are installed with black mushrooms, bamboo shoots, soy sauce and some sugar and simmered for about 10 min. Finally, some cornstarch mixed with tablespoon water and sesame oil are added and tossed gently. It can be served with white rice and vegetables (Figs. 7.12 and 7.13) (Ligia, 2020).

FIGURE 7.12

Sea cucumber in sauce with rice and broccoli.

Credit: Mohamed Mohsen.

FIGURE 7.13

Sea cucumber dish with sauce and broccoli.

Credit: Kent Wang. Licensed under CC BY-SA 2.0 via Wikimedia commons.

7.2.6 Braised sea cucumber with scallions

After rehydration, sea cucumbers are cut into pieces and set aside. Next, slices of mushrooms and bamboo shoots are fried. Then, pieces of sea cucumbers and sauce (dark soy sauce, oyster sauce, light soy sauce, sugar and some water) are added and boiled until the mixture becomes thick. Finally, pieces of scallions are added, and the mixture is fried before serving (Fig. 7.14) (He, 2020).

FIGURE 7.14

Sea cucumber cooked with scallions in a sauce made by soy sauce and sesame oil.

Credit: Zheng Zhou. Licensed under CC BY-SA 4 via Wikimedia commons.

References

Ahmed, M., 2009. Morphological, Ecological and Molecular Examination of the Sea Cucumber Species along the Red Seacoast of Egypt and Gulf of Aqaba. The University of Hull.

Antal, T., 2015. Comparative study of three drying methods: freeze, hot air assisted freeze and infrared-assisted freeze modes. Agron. Res. 13, 863–878.

Bai, Y., Luan, Z., 2018. The effect of high-pulsed electric field pretreatment on vacuum freeze drying of sea cucumber. Int. J. Appl. Electromagn. Mech. 57, 247–256. https://doi.org/10.3233/JAE-180009.

Bai, Y., Qu, M., Luan, Z., Li, X., Yang, Y., 2013. Electrohydrodynamic drying of sea cucumber (*Stichopus japonicus*). LWT - Food Sci. Technol. 54, 570–576. https://doi.org/10.1016/j.lwt.2013.06.026.

Bai, Y., Yang, Y., Huang, Q., 2012. Combined electrohydrodynamic (ehd) and vacuum freeze drying of sea cucumber. Dry. Technol. 30, 1051–1055. https://doi.org/10.1080/07373937.2012.663435.

Bilgin, Ş., Öztürk Tanrikulu, H., 2018. The changes in chemical composition of *Holothuria tubulosa* (Gmelin, 1788) with ambient-drying and oven-drying methods. Food Sci. Nutr. 6, 1456–1461. https://doi.org/10.1002/fsn3.703.

Chong, N.V.W., Pindi, W., Chye, F.Y., Shaarani, S.M., Lee, J.S., 2015. Effects of drying methods on the quality assessments of dried sea cucumbers from Sabah. Int. J. Nov. Res. Life Sci. 4, 49–64.

Defraeye, T., Martynenko, A., 2018. Future perspectives for electrohydrodynamic drying of biomaterials. Dry. Technol. https://doi.org/10.1080/07373937.2017.1326130.

Drouzas, A.E., Schubert, H., 1996. Microwave application in vacuum drying of fruits. J. Food Eng. 28, 203–209. https://doi.org/10.1016/0260-8774(95)00040-2.

Duan, X., Zhang, M., Li, X., Mujumdar, A.S., 2008. Microwave freeze drying of sea cucumber coated with nanoscale silver. Dry. Technol. 26, 413–419. https://doi.org/10.1080/07373930801929136.

Duan, X., Zhang, M., Mujumdar, A.S., 2007. Study on a combination drying technique of sea cucumber. Dry. Technol. 25, 2011−2019. https://doi.org/10.1080/07373930701728497.

Duan, X., Zhang, M., Mujumdar, A.S., Huang, L.L., Wang, H., 2010a. A novel dielectric drying method of sea cucumber. Int. J. Food Sci. Technol. 45, 2538−2545. https://doi.org/10.1111/j.1365-2621.2010.02419.x.

Duan, X., Zhang, M., Mujumdar, A.S., Wang, S., 2010b. Microwave freeze drying of sea cucumber (*Stichopus japonicus*). J. Food Eng. 96, 491−497. https://doi.org/10.1016/j.jfoodeng.2009.08.031.

Duan, X., Zhang, M., Mujumdar, S., Wang, R., 2010. Trends in microwave-assisted freeze drying of foods. Dry. Technol. 28, 444−453. https://doi.org/10.1080/07373931003609666.

Hayashi, H., 1989. Drying technologies of foods - their history and future. Dry. Technol. 7, 315−369. https://doi.org/10.1080/07373938908916590.

He, V., 2020. The Braised Sea Cucumber with Scallion. URL: https://misschinesefood.com/the-braised-sea-cucumber-with-scallion/.

Hernández, O.A., Pabón, E.A., Montoya, O.J.C., Duran, E.C., Narváez, R.O.C., Forero, A.R., 2017. Sea cucumber (*Isostichopus* sp. aff badionotus) dry-salting protocol design. Nat. Resour. 8, 278−289.

Jenny, E., Heather, B., 2016. Vegan 101: A Vegan Cookbook: Learn to Cook Plant-Based Meals that Satisfy Everyone. Arcas Publishing.

Jianrong, L., Xyuepeng, L., Ping, Y., Jie, Y., Zhibin, X., 2007. Method for Preparing Sea Cucumber.

Lavitra, T., Rachelle, D., Rasolofonirina, R., Jangoux, M., Eeckhaut, I., 2008. Processing and marketing of holothurians in the Toliara region, southwestern Madagascar. Beche-de-Mer Bull. 28, 24−33.

Li, C., Li, H., Guo, S., Li, X., Zhu, X., 2018. Evaluation of processing methods on the nutritional quality of sea cucumber (*Apostichopus japonicus* Selenka). J. Aquat. Food Prod. Technol. 27, 406−417. https://doi.org/10.1080/10498850.2013.803271.

Li, M., Qi, Y., Mu, L., Li, Z., Zhao, Q., Sun, J., Jiang, Q., 2019. Effects of processing method on chemical compositions and nutritional quality of ready-to-eat sea cucumber (*Apostichopus japonicus*). Food Sci. Nutr. 7, 755−763. https://doi.org/10.1002/fsn3.921.

Liew, A., 2015. Braised Sea Cucumber with Fish Maw. URL: https://www.huangkitchen.com/braised-sea-cucumber-with-fish-maw/. (Accessed 10 September 2020).

Ligia, 2020. Braised Sea Cucumber and Japanese Mushroom with Seasonal Vegetable. URL: https://www.pscha.org/2020/06/16/braised-sea-cucumber-and-japanese-mushroom-with-seasonal-vegetable/. (Accessed 10 September 2020).

Liu, Q., Sun, J., Pang, Y., Jia, Z., 2016. Optimization of processing technology of instant sea cucumber with fuzzy mathematic comprehensive evaluation by response surface methodology and exploration on nutritional value of instant sea cucumber. Food Sci. Technol. Res. 22, 583−593. https://doi.org/10.3136/fstr.22.583.

Meng, S., Zhang, H., Sun, J., Liu, Q., 2017. Study on nutritional value of instant sea cucumber processed by the new processing approach. Am. J. Biochem. Biotechnol. 13, 51−57. https://doi.org/10.3844/ajbbsp.2017.51.57.

Moon, J.H., Kim, M.J., Chung, D.H., Pan, C.H., Yoon, W.B., 2014. Drying characteristics of sea cucumber (*Stichopus japonicus* Selenka) using far infrared radiation drying and hot air drying. J. Food Process. Preserv. 38, 1534−1546. https://doi.org/10.1111/jfpp.12113.

Moon, J.H., Yoon, W.B., 2015. Size dependence of the salting process for dry salted sea cucumber (*Stichopus japonicus*). J. Food Eng. 170, 170—178. https://doi.org/10.1016/j.jfoodeng.2015.09.028.

Öztürk, F., Gündüz, H., 2018. The effect of different drying methods on chemical composition, fatty acid, and amino acid profiles of sea cucumber (*Holothuria tubulosa* Gmelin, 1791). J. Food Process. Preserv. 42 https://doi.org/10.1111/jfpp.13723.

Pan, S., Wu, S., Yao, D., 2014. Inhibition of shrinkage of sea cucumber *Stichopus japonicus* during canning using sodium phytate and sodium tripolyphosphate. Food Sci. Technol. Res. 20, 279—282. https://doi.org/10.3136/fstr.20.279.

Qian, Z., Guochen, Z., Gang, M., Yue, L., 2012. Freeze and microwave vacuum combination drying technique of sea cucumber. Int. J. Agric. Biol. Eng. 5 https://doi.org/10.3965/j.ijabe.20120503.00?.

Rasolofonirina, R., Mara, E., And, M., 2004. Sea Cucumber Fishery and Mariculture in Madagascar, a Case Study of Toliara, Southwest Madagascar. Adv. sea cucumber Aquac. Manag. FAO Fish. Tech. Pap. 463, pp. 133—149.

Ravinesh Ram, P.C., 2014. Effects of harvest and post-harvest processing methods on quality of beche-de-mer in Fiji Islands. J. Mar. Sci. Res. Dev. 04, 1—6. https://doi.org/10.4172/2155-9910.1000153.

Roveena Vandana Chand, R.R., 2014. Effects of processing methods on the value of bêche-de-mer from the Fiji Islands. J. Mar. Sci. Res. Dev. 04, 3. https://doi.org/10.4172/2155-9910.1000152.

Schoppe, S., 2000. Sea cucumber fishery in the Philippines. SPC Beche-de-mer Inf. Bull. 13, 1—12.

Subaldo, M.C., 2011. Gleaning, drying and marketing practices of sea cucumber in Davao Del Sur, Philippines. JPAIR Multidiscip. Res. 6 https://doi.org/10.7719/jpair.v6i1.140.

Telahigue, K., Hajji, T., Imen, R., Sahbi, O., Cafsi, M. El, 2014. Effects of drying methods on the chemical composition of the sea cucumber *Holothuria forskali*. Open Food Sci. J. 8, 1—8. https://doi.org/10.2174/1874256401408010001.

Vaipulu, S.'E.U., 2009. Design a Prototype Solar Dryer for Drying Sea Cucumber. University of Southern Queensland.

Veneracion, C., 2018. Chinese-style Braised Sea Cucumbers and Shiitake Mushrooms. URL: https://casaveneracion.com/chinese-style-braised-sea-cucumbers-and-shiitake-mushrooms/. (Accessed 10 September 2020).

Zhang, G.C., Mu, G., Wang, J.D., Zhang, Q., 2012. Microwave vacuum drying process of semi dry salted sea cucumber. J. Dalian Ocean Univ. 2.

Zhang, M., Jiang, H., Lim, R.X., 2010. Recent developments in microwave-assisted drying of vegetables, fruits, and aquatic products-drying kinetics and quality considerations. Dry. Technol. 28, 1307—1316. https://doi.org/10.1080/07373937.2010.524591.

Developing sea cucumbers aquaculture in the Middle East: a perspective

8.1 Background

Sea cucumbers, also known as holothurians, are epibenthic echinoderms that have elongated bodies. The mouth is at one end and encircled by extended podia known as tentacles, which vary in shape and number, and the anus is at the other end (Chapter 1). The thickness of the body wall determines the commercial importance of the sea cucumber species. The body wall constitutes of epidermis and dermis layers; the dermis layer constitutes the more significant part of the body walls in which the ossicles are buried. These ossicles have a taxonomic function in the holothuroids. The lumen of the body is filled with the coelomic fluid in which the coelomocytes float. The coelomocytes have many functions, including immune defence, nutrient transport, gas exchange and waste execration. Inside the body cavity, there is an intestine, which starts from the mouth, pharynx, oesophagus, stomach, intestine, rectum and finally ends by the cloaca. Two branches of the respiratory tree extend in the body cavity in the orders Aspidochirota, Dendrochirota and Molpadonia, which serve as a respiratory function. The body cavity also contains gonads. Sea cucumbers feel their surroundings through nerve plexus permeated in the body wall. Most of the holothurians are dioecious; the sex is indistinguishable externally (Chapter 2). Sea cucumbers reproduce sexually, while some of them reproduce asexually by fission. When threatened, sea cucumbers explode their internal organs, and they can regenerate them, which is considered an extraordinary aspect of the holothurians (Chapter 3).

Sea cucumbers have received significant attention from researchers worldwide for their nutritional and medicinal values and so-called jewels of the seabed (Kiew and Don, 2012). Sea cucumbers contain many highly valued compounds, and their bioactive compounds are linked to multiple biological effects, such as anti-cancer, anti-inflammatory, anti-angiogenic, anti-coagulant, anti-hypertension, anti-inflammatory, anti-microbial, anti-oxidant, anti-thrombotic, wound healing

Sea Cucumbers. https://doi.org/10.1016/B978-0-12-824377-0.00003-7

and fertility improving (Chapters 4 and 5). Furthermore, sea cucumbers play a crucial role in the marine ecosystem. Sea cucumbers improve sediment characteristics and their associated fauna. Also, sea cucumbers improve water chemistry and remove aquaculture wastes (Chapters 4 and 5).

Sea cucumber resources have a long history of exploitation, driven by the high value of sea cucumbers in Asian markets. Consequently, overfishing has become a worldwide phenomenon and accompanied by poor fisheries management, which severely decreased commercial sea cucumber populations worldwide. As a result, artificial reproduction of sea cucumber received much attention to satisfy demand and assuage stock decline. Artificial production of sea cucumber species started with the sea cucumber *Apostichopus japonicus*. Then, the production methods were adopted to other species, such as the sea cucumber *Holothuria scabra*, with small adjustments. Investments and considerable research effort in the sandfish *H. scabra* allowed the production extensively at commercial scale and have led to some significant advances (Chen, 2004; Eriksson et al., 2012; Robinson, 2013).

Following the pioneering work on the hatchery and culture of the sea cucumber *A. japonicus*, the industrial farming of *A. japonicus* has developed in several Asian countries, especially in China (Chen, 2004). Artificial breeding of *A. japonicus* in China has witnessed rapid development since the 1980s as a result of the improvement and the development of breeding programmes (Lovatelli et al., 2004). The broodstock was collected from the wild and almost directly induced to spawn. After that, facilities were established to allow gonads to attain full maturity. Also, favourable indoor conditions were maintained (e.g. temperature, darkness, feeding and the composition and quality of food) in order to gain an extension of spawning period, gain high numbers of individual spawning and produce large-scale juveniles. Furthermore, technological improvements enabled to induce spawning using a variety of methods and improved the hatching rate (Xilin, 2004).

The growth of the larvae was improved by providing a balanced diet. Series of studies have been conducted to improve the survival rate and the growth of the juvenile in the nursing stage through developing a formulated diet (Gao et al., 2011; Xilin, 2004; Yu et al., 2016; Zhang et al., 2010). Furthermore, high-quality seawater maintenance (e.g. temperature, pH, salinity, ammonia, dissolved oxygen, heavy metal concentration and turbidity) was a prerequisite of successful breeding (Xilin, 2004). Moreover, substrates were further developed to increase the success of settlement of the juvenile (Jiang et al., 2015; Li et al., 2010; Xilin, 2004). Cultivating the benthic diatoms on the settling bases was widely used. Then, hatcheries started providing the food after settling the juvenile instead of culturing the benthic diatoms on the settling bases, increasing the survival rate (Xilin, 2004; Zhang et al., 2015).

Furthermore, effective pesticides were developed to kill the predators (e.g. copepods), which were found in the rearing tanks. In recent years, the industrial indoor breeding was further developed to support the seedlings for outdoor culture facilities. Technological improvements allowed the control of high water quality as well as disease prevention and treatment, using a filtration system, UV sterilisation, oxygen pumping and heating. Also, research units are basic units on the site, providing technical support. Furthermore, in order to produce cost-effective facilities for sea cucumber aquaculture, wind and solar energy are utilised to transport seawater, heating and generate electricity (Zhang et al., 2015).

The artificial breeding and rearing of seedlings started in ponds in offshore areas using former shrimp ponds then using newly built ponds. Next, the farming methods developed from indoor tanks to sea ranching and suspended culture, which contributed significantly to raising the total production of *A. japonicus*, which has become an essential industry after years of developments. Furthermore, sea ranching was developed as a feasible and eco-friendly farming method. It is established in a suitable condition for sea cucumber growth, where natural food is available. Therefore, this method offers a high growth rate, less incidence of disease and less production cost (Chen, 2004). As a result, the sea cucumber farming industry in China has been developing rapidly in order to meet the increasing market demand. According to data from the China Fishery Statistical Yearbook, the yield, mariculture area and quantity of sea cucumber seedlings have significantly increased in recent years.

Despite the growing numbers of studies focussing on sea cucumbers worldwide, sea cucumbers in the Middle East have not received much attention. Research on sea cucumber aquaculture has been conducted on some sea cucumber species, including the sea cucumbers *Holothuria tubulosa* (Günay et al., 2018; Rakaj et al., 2018; Tolon et al., 2017; Toscano and Cirino, 2018), *Holothuria arenicola* (Razek et al., 2012), *Holopthuria poli* (Rakaj et al., 2019; Toscano and Cirino, 2018), *H. scabra* (Al Rashdi et al., 2012; Dabbagh and Sedaghat, 2012), *Holothuria leucospilota* (Dabbagh et al., 2011; Soltani et al., 2010) and *Actinopyga mauritiana* (Gabr et al., 2005) (see Chapters 4 and 5 for further details). However, commercial aquaculture still needs to be expanded on a large scale. Also, some problems were encountered with the breeding process. One problem was because of the copepods' infestation during larval rearing of sea cucumber *H. scabra* in Oman, which resulted in mass mortality during few days post fertilisation (Al Rashdi et al., 2012). Also, in Iran, the high water temperature of over 30°C year-round and the high density were the major problems encountered with the production of the sea cucumber *H. scabra* (Dabbagh and Sedaghat, 2012), and further research is required to improve the observed low growth rate (Dabbagh and Sedaghat, 2012). These previous experiences should be considered to improve the aquaculture process of the intended cultured species.

8.2 Potential species for aquaculture

The sea cucumber *H. scabra* is the most intensively studied species in the temperate region for the monoculture pond system. Also, researchers reported many successful breeding trials for sea cucumber species that could be bred in the captivity (Chapter 6).

The sea cucumber *H. scabra* was chosen for aquaculture because it has many features: a high price in the Chinese markets, can be produced in hatcheries, capable of rapid growth (1—3 g/day) under suitable conditions, can reach maturity at less than 1 year of age and can be spawned year-round under tropical conditions (Pitt and Duy, 2004; Tuwo and Tresnati, 2015; Watanabe et al., 2014). Also, the sea cucumbers *H. leucospilota* and *Holothuria arguinensis* have a high price, and they can be produced in hatcheries, but their aquaculture on a large scale needs further exploration.

8.3 Potential areas for sea cucumber mariculture

The management of sea cucumber aquaculture is similar, in some respects, to other species aquaculture. Indeed, sea cucumbers aquaculture has superior features, such as harvesting ease and low feeding cost. Sea cucumber aquaculture requires monitoring of the bottom environment together with supplying high water quality and preventing diseases. The bottom of the farm should be constructed in a way that is suitable to the benthic lifestyle of the sea cucumber. Therefore, an improvement of the biological knowledge of the target species is essential. For successful sea cucumber marine ranching, a survey of water quality, seafloor condition and the distribution of other marine organisms will help to determine the appropriate location. Following the choice of a suitable area with high water quality (Fig. 8.1), the seafloor often needs to be enhanced to suit the sea cucumber species of choice (Figs. 8.2 and 8.3). Overall, sea cucumber mariculture can be successful in coastal areas with high water quality, which can be adapted for the benthic lifestyle of the sea cucumber species.

FIGURE 8.1

Sea cucumbers farm in a coastal area in China.

Credit: Chenggang Lin.

FIGURE 8.2

Enhancing the seafloor by providing artificial substrates for sea cucumbers.

Credit: Shilin Liu.

FIGURE 8.3

Deployment of rocky substrates in sea cucumber ponds.

Credit: Mohamed Mohsen.

8.4 Potential models for sea cucumber mariculture

In China, the suspended culture system of sea cucumbers has been developed. Various models are used in the suspended culture of the sea cucumbers with less production cost and a high harvesting rate. These models were adopted from kelp, scallops and abalone systems (Fig. 8.4). The cage is designed to suit the feeding behaviour of the sea cucumber, which provides protection and source of growing natural food. It can be made from plastic with a top open or mesh with a zipper. Additional food can be added once a week, or another species can be added with sea cucumber as a food source for better growing sea cucumber. In this case, sea cucumber can be added on the bottom culture plate. *A. japonicus* reached a maximum

FIGURE 8.4

Stack cage systems adopted from traditional abalone culture cages.

Credit: Shilin Liu.

growth rate of 0.34%/d using the multispecies cage system. Stocking densities were reported as 22.3 ind./m^2 with feed and 14.1 ind./m^2 without feed. Also, farmers should carefully monitor the water quality parameters in the area of the suspended culture (Yuan et al., 2008; Zhang et al., 2015).

Moreover, marine ranching sea cucumbers in China has become a widespread practice. Sea ranching is exploiting a specific area of the ocean to establish a suitable environment for farming sea cucumbers (Fig. 8.5). Sea cucumber can be cultured in

FIGURE 8.5

Marine ranching site in China.

Credit: Chenggang Lin.

suitable areas when the food is available. For instance, a rocky or muddy-sandy site with low currents and rich in seagrass is favourable for the successful sea ranching of *A. japonicus*. Adding artificial reefs can create suitable habitat, providing shelter and food, for sea cucumber (e.g. rocks, seaweed beads and oyster shell reef) (Fig. 8.6). Oyster shell reefs can be used in areas with fine silt to provide a suitable substrate for the sea cucumber. This type of reef consists of a mesh containing 20−75 kg of oyster shells. Juveniles of 3 mm can be initially stocked in the sea ranching area with a stocking density of 3−15 m^2 (Han et al., 2016; Zhang et al., 2015).

Furthermore, the development of the integrated aquaculture system of sea cucumbers with other taxa is of a great potential value given the deposit-feeding behaviour of sea cucumbers and their position in the food chain. Sea cucumbers were successfully co-cultured with abalone, fish, shrimp, mussel, scallop, seaweed and sea urchin. Also, sea cucumbers were successfully integrated under or in net cages, in ponds and integrated into the recirculated aquaculture system (see Chapter 6 for further details). Although these successful co-culture trials were mainly conducted using the sea cucumbers *A. japonicus* and *H. scabra*, it increases the potential of using other sea cucumber species likewise, which worth further exploration.

FIGURE 8.6

Installing artificial reefs in marine ranching site in China.

Credit: Mohamed Mohsen.

8.5 Sea cucumber marketing

Although sea cucumbers are dried and prepared for exportation, their local marketing opportunities need to be explored and encouraged, using processing and drying techniques, such as ready-to-eat and canned sea cucumbers (Chapter 7; Fig. 8.7). Improving the sea cucumber industry needs well-conceived strategies that would improve primary processing, sustain supply and maximise returns (Perez and Brown, 2012). Also, a better understanding of the trade, value and market preferences is essential for management strategies (Rahman and Yusoff, 2017).

FIGURE 8.7

Sea cucumbers marketing in China.

Credit: Mohamed Mohsen.

8.6 Research plan

Researchers, farmers, governments and enterprises should fully cooperate in pursuing the goal of successful sea cucumber aquaculture. Following previous experiences (Dabbagh and Sedaghat, 2012; Eriksson et al., 2012; Giraspy and Walsalam, 2010; Han et al., 2016; Lawrence et al., 2004; Lovatelli et al., 2004; Mercier et al., 2005; Purcell, 2004; Purcell et al., 2012; Rasolofonirina et al., 2004; Renbo and Yuan, 2004), the key components of the action plan are as follows:

- Raise the awareness of fishers to respect fishery legislations.
- Research on the biological and ecological aspects of sea cucumbers and potential species for aquaculture.
- Research on reproductive biology, size at first maturity and estimation of growth rate.
- Implementation of regulations for stock protection and limit the depletion of the stocks.
- Accurate evaluation of stocks and provide potential sites for broodstock collection.
- Feasibility study of the venture and reduce the labour cost.
- Strong institutional support as well as donor fund to invest in hatchery facilities.
- Research on broodstock management for successful spawning.
- Research on conditioning the broodstock to avoid disruption of the wild population.

- Refine and simplifying the customised hatchery and larval rearing protocols to maximise commercial mass production.
- Developing optimal larval rearing protocol as well as water quality management and disease control.
- Research on diets and stocking density of the larvae stage.
- Provide agencies for technical advice and training.
- Research regarding the best outdoor grow-out techniques and aquaculture facility construction, optimal stocking density and substrates and provide suitable locations to grow the juveniles.
- Research on sea cucumbers co-culture feasibility.
- Developing optimal releasing strategies for hatchery-produced sea cucumbers.
- Knowledge of suitable releasing sites through genetic structure assessment.
- Assess the genetic diversity and gene flow among populations.
- Improve processing and storage methods of sea cucumbers to preserve their nutritional composition and facilitate the marketing process.
- Information on international marketing and regulation as well as research on the potential of local marketing.

References

Al Rashdi, K.M., Eeckhaut, I., Claereboudt, M.R., 2012. A manual on hatchery of sea cucumber *Holothuria scabra* in the Sultanate of Oman. Minist. Agric. Fish. Wealth, Aquac. Centre, Muscat, Sultanate Oman, p. 27.

Chen, J., 2004. Present status and prospects of sea cucumber industry in China. Adv. Sea Cucumber Aquacult. Manag. 463, 25—38.

Dabbagh, A.-R., Sedaghat, M.R., 2012. Breeding and rearing of the sea cucumber *Holothuria scabra* in Iran. SPC Beche-de-mer Inf. Bull. 49—52.

Dabbagh, A.-R., Sedaghat, M.R., Rameshi, H., Kamrani, E., 2011. Breeding and larval rearing of the sea cucumber *Holothuria leucospilota* Brandt (*Holothuria vegabunda* Selenka) from the northern Persian Gulf, Iran. SPC Beche-de-mer Inf. Bull. 35—38.

Eriksson, H., Robinson, G., Slater, M.J., Troell, M., 2012. Sea cucumber aquaculture in the western Indian Ocean: challenges for sustainable livelihood and stock improvement. Ambio. https://doi.org/10.1007/s13280-011-0195-8.

Gabr, H.R., Ahmed, A.I., Hanafy, M.H., Lawrence, A.J., Ahmed, M.I., El-Etreby, S.G., 2005. Mariculture of sea cucumber in the Red Sea -the Egyptian experience. FAO Fish. Aquacult. Tech. Pap. 373—384.

Gao, Q.F., Wang, Y., Dong, S., Sun, Z., Wang, F., 2011. Absorption of different food sources by sea cucumber *Apostichopus japonicus* (Selenka) (Echinodermata: Holothuroidea): evidence from carbon stable isotope. Aquaculture 319, 272—276. https://doi.org/10.1016/j.aquaculture.2011.06.051.

Giraspy, D., Walsalam, I., 2010. Aquaculture potential of the tropical sea cucumbers *Holothuria scabra* and *H. lessoni* in the Indo-Pacific region. SPC Beche-de-mer Inf. Bull. 30, 29—32.

Günay, D., Tolon, M.T., Emiroğlu, D., 2018. Effects of various stocking densities on hatching and survival rates of sea cucumber *Holothuria tubulosa* eggs (Gmelin, 1788). Ege J. Fish. Aquat. Sci. 35, 381–386. https://doi.org/10.12714/egejfas.2018.35.4.03.

Han, Q., Keesing, J.K., Liu, D., 2016. A review of sea cucumber aquaculture, ranching, and stock enhancement in China. Rev. Fish. Sci. Aquacult. 24 (4), 326–341. https://doi.org/10.1080/23308249.2016.1193472.

Jiang, S., Dong, S., Gao, Q., Ren, Y., Wang, F., 2015. Effects of water depth and substrate color on the growth and body color of the red sea cucumber, *Apostichopus japonicus*. Chin. J. Oceanol. Limnol. 33, 616–623. https://doi.org/10.1007/s00343-015-4178-7.

Kiew, P.L., Don, M.M., 2012. Jewel of the seabed: sea cucumbers as nutritional and drug candidates. Int. J. Food Sci. Nutr. https://doi.org/10.3109/09637486.2011.641944.

Lawrence, A.J., Ahmed, M., Hanafy, M., Gabr, H., Ibrahim, A., Gab-Alla, A.-F., 2004. Status of the sea cucumber fishery in the Red Sea - the Egyptian experience. Adv. Sea Cucumber Aquacult. Manag. FAO Fish. Tech. Pap. 463.

Li, L., Li, Q., Kong, L., 2010. The effect of different substrates on larvae settlement in sea cucumber, *Apostichopus japonicus* Selenka. J. World Aquacult. Soc. 41, 123–130. https://doi.org/10.1111/j.1749-7345.2009.00341.x.

Lovatelli, A., Conand, C., Purcell, S., Uthicke, S., Hamel, J.-F., Mercier, A., 2004. Advances in sea cucumber aquaculture and management. FAO Fish. Tech. Pap. 463, 1–440.

Mercier, A., Hidalgo, R., Hamel, J., 2005. Aquaculture of the Galapagos sea cucumber, *Isostichopus fuscus*. Fao Fish. Tech. Pap. 347–358.

Perez, M.L., Brown, E.O., 2012. Market potential and challenges for expanding the production of sea cucumber in South-East Asia. Asia-Pac. Trop. Sea Cucumber Aquacult. ACIAR Proc. 136, 177–188.

Pitt, R., Duy, N.D.Q., 2004. Breeding and rearing of the sea cucumber *Holothuria scabra* in Vietnam. In: Advances in Sea Cucumber Aquaculture and Management, pp. 333–346.

Purcell, S., 2004. Criteria for release strategies and evaluating the restocking of sea cucumbers. Adv. Sea Cucumber Aquacult. Manag. 181–191.

Purcell, S.W., Hair, C.A., Mills, D.J., 2012. Sea cucumber culture, farming and sea ranching in the tropics: progress, problems and opportunities. Aquaculture. https://doi.org/10.1016/j.aquaculture.2012.08.053.

Rahman, M.A., Yusoff, F.M., 2017. Sea cucumber fisheries: market potential, trade, utilisation and challenges for expanding the production in the South-East Asia. Int. J. Adv. Chem. Eng. Biol. Sci. 4 https://doi.org/10.15242/ijacebs.er0117033.

Rakaj, A., Fianchini, A., Boncagni, P., Lovatelli, A., Scardi, M., Cataudella, S., 2018. Spawning and rearing of *Holothuria tubulosa*: a new candidate for aquaculture in the Mediterranean region. Aquacult. Res. 49, 557–568. https://doi.org/10.1111/are.13487.

Rakaj, A., Fianchini, A., Boncagni, P., Scardi, M., Cataudella, S., 2019. Artificial reproduction of *Holothuria polii*: a new candidate for aquaculture. Aquaculture 498, 444–453. https://doi.org/10.1016/j.aquaculture.2018.08.060.

Rasolofonirina, R., Mara, E., And, M., 2004. Sea cucumber fishery and mariculture in Madagascar, a case study of Toliara, southwest Madagascar. Adv. Sea Cucumber Aquacult. Manag. FAO Fish. Tech. Pap. 463, 133–149.

Razek, F.A.A., Rahman, S.H.A., Moussa, R.M., Mena, M.H., El-Gamal, M.M., 2012. Captive spawning of *Holothuria arenicola* (Semper, 1868) from Egyptian Mediterranean Coast. Asian J. Bio. Sci. 5, 425–431. https://doi.org/10.3923/ajbs.2012.425.431.

Renbo, W., Yuan, C., 2004. Breeding and culture of the sea cucumber, *Apostichopus japonicus*, Liao. In: Advances in Sea Cucumber Aquaculture and Management, pp. 277–286.

Robinson, G., 2013. A bright future for sandfish aquaculture. World Aquacult. Soc. Mag. 18—24.

Soltani, M., Radkhah, K., Mortazavi, M.S., Gharibniya, M., 2010. Early development of the sea cucumber *Holothuria leucospilota*. Res. J. Anim. Sci. 4, 72—76. https://doi.org/10.3923/rjnasci.2010.72.76.

Tolon, T., Emiroğlu, D., Günay, D., Hancı, B., 2017. Effect of stocking density on growth performance of juvenile sea cucumber *Holothuria tubulosa* (Gmelin, 1788). Aquacult. Res. 48, 4124—4131. https://doi.org/10.1111/are.13232.

Toscano, A., Cirino, P., 2018. First evidence of artificial fission in two Mediterranean species of holothurians: *Holothuria tubulosa* and *Holothuria polii*. Turk. J. Fish. Aquat. Sci. 18, 1141—1145. https://doi.org/10.4194/1303-2712-v18_10_01.

Tuwo, A., Tresnati, J., 2015. Sea cucumber farming in Southeast Asia (Malaysia, Philippines, Indonesia, Vietnam). In: Echinoderm Aquaculture, pp. 331—352. https://doi.org/10.1002/9781119005810.ch15.

Watanabe, S., Sumbing, J.G., Lebata-Ramos, M.J.H., 2014. Growth pattern of the tropical sea cucumber, *Holothuria scabra*, under captivity. Jpn. Agric. Res. Q. 48, 457—464. https://doi.org/10.6090/jarq.48.457.

Xilin, S., 2004. The progress and prospects of studies on artificial propagation and culture of the sea cucumber, *Apostichopus japonicus*. In: Advances in Sea Cucumber Aquaculture and Management. FAO, Rome (Italy), pp. 273—276.

Yu, H., Gao, Q., Dong, S., Zhou, J., Ye, Z., Lan, Y., 2016. Effects of dietary n-3 highly unsaturated fatty acids (HUFAs) on growth, fatty acid profiles, antioxidant capacity and immunity of sea cucumber *Apostichopus japonicus* (Selenka). Fish. Shellfish Immunol. 54, 211—219. https://doi.org/10.1016/j.fsi.2016.04.013.

Yuan, X.T., Yang, H.S., Zhou, Y., Mao, Y.Z., Xu, Q., Wang, L.L., 2008. Bioremediation potential of *Apostichopus japonicus* (Selenka) in coastal bivalve suspension aquaculture system. Chin. J. Appl. Ecol. 19, 866—872.

Zhang, L., Song, X., Hamel, J.F., Mercier, A., 2015. Aquaculture, stock enhancement, and restocking. Dev. Aquacult. Fish. Sci. https://doi.org/10.1016/B978-0-12-799953-1.00016-7.

Zhang, Q., Ma, H., Mai, K., Zhang, W., Liufu, Z., Xu, W., 2010. Interaction of dietary *Bacillus subtilis* and fructooligosaccharide on the growth performance, non-specific immunity of sea cucumber, *Apostichopus japonicus*. Fish. Shellfish Immunol. 29, 204—211. https://doi.org/10.1016/j.fsi.2010.03.009.

Index

'Note: Page numbers followed by "f" indicate figures and "t" indicate tables.'

A

Actinopyga agassizii, 6—7, 7f
Actinopyga caerulea, 4, 6f
Actinopyga mauritiana, 19—21, 79—80, 79f
Aegean Sea, 65, 68, 86
Aestivation, 42—43, 44f
Algae path, 132—133
Algal bloom cleaning, 48—49
Anti-bacterial properties, 89—90
Anti-fungal properties, 90—91
Anti-microbial properties, 114—115
Anti-oxidant properties, 117
Anti-tumour properties, 115—117
Anti-viral properties, 117
Anus, 4
Apodida, 1
Apostichopus californicus, 4, 5f, 133—134
Apostichopus japonicus, 19—21, 21f, 41, 41f,
  48—49, 174
Aquaculture, 110—111, 175—176
  asexual reproduction, 83—84
  sexual reproduction, 84
  waste bioremediation, 49—50
Arachidonic acid, 111—114, 111f
Artificial breeding, 174
Artificial fission, 34
Asexual reproduction, 31—32, 83—84
Aspidochirotida, 1, 38
Asterina pectinifera, 53, 53f
Australostichopus mollis, 38—39

B

Benthic organisms, 50—52
Benthic oxygen regeneration, 48
Binary fission, 31—32, 32f
Bioactive compounds, 9, 10f, 10t—13t, 88—89
Bioremediation, 49—50
Bioturbation, 47—48, 47f
Body walls, 19, 20f, 91
  epidermis and dermis layers, 19, 173
  polysaccharides, 9
  properties, 19
  thickness of, 1
Bohadschia argus, 69, 70f
Bohadschia marmorata, 82, 83f
Bohadschia ocellate, 1, 2f
Bohadschia vitiensis, 82, 83f

C

Broodstock, 84, 110
  collection, 127—128, 127t
  conditioning, 130—131, 131t

Calcareous ring, 21—23, 23f
Calcium carbonate ($CaCO_3$), 49
Canned sea cucumbers, 164, 164f
Circulatory system
  coelomic fluid and coelomocytes, 24, 24f
  haemal system, 23
Cladolabes schmeltzii, 27, 32—34, 32f
Co-culture, 146, 147t—149t
Coelomic fluid, 24, 24f
Coelomocytes, 24, 24f
Connective tissue, 19
Cooking, 165f
  braised sea cucumbers, 165, 165f
    scallions, 168, 169f
    vegetables, 167, 168f
  deep-fried sea cucumber, 166, 167f
  mushrooms, 166, 167f
  salad, 166, 166f
Cucumaria frondosa, 29—31, 30f
Cucumaria miniata, 4, 5f, 133—134
Cuvierian tubules, 37—38, 38f, 63

D

Dactylochirotida, 1
Dendrochirotida, 1, 38
Digestive system
  histology, 26, 26f
  morphology, 24—25, 25f
Docosahexaenoic acid, 87
Dried sea cucumbers, 8—9, 8f
Drying method, 132
Dry-salting method, 162

E

Early growth response protein 1 (Egr1), 43
Echinacoside, 91, 92f
Ecological values
  sediment characteristics, 85—86
  surrounding environment, 86
Eicosapentaenoic acid (EPA), 87
Elasipodida, 1
Electrohydrodynamic (EHD) drying, 161

Embryo development, 137–143
Endoskeleton, 19–21
Enterocytes, 26
Environmental factors, 143–144
Eupentacta quinquesemita, 1, 3f
Eutrophication, 48
Evisceration, 37–38, 37f–38f

F
Farming, 149–150, 150f
Fertilisation, 135–137, 136f, 137t
Freeze-drying, 160–161
Fucosylated chondroitin sulphate, 91, 93f

G
Gadoleic acid, 111–114, 112f
Gametogenesis, 131
Gene flow, 43, 44f
Genetic diversity, 43, 45
Glutamic acids, 9
Glycine, 9
Gonadal development, 128, 128t–129t, 130f
Grow-out methods
    co-culture, 146, 147t–149t
    earthen ponds, 145
    sea pens, 145
    suspended culture, 145–146
Gulf of Aqaba, 62

H
Haemal system, 23
Hawaiian Archipelago, 46
Heneicosanoic acid, 111–114, 113f
High-pulsed electric field (HPEF), 161
Holothuria arenicola, 69, 69f, 109
Holothuria arguinensis, 65–66, 65f
Holothuria atra, 46, 80–81, 80f
Holothuria bacilla, 109
Holothuria edulis, 81–82, 82f
Holothuria forskali, 37–38, 38f, 69, 70f
Holothuria fuscogilva, 77, 77f
Holothuria glaberrima, 30f
Holothuria hilla, 107, 107f
Holothuria impatiens, 108, 108f
Holothuria leucospilota, 81, 81f, 106, 106f, 140f
Holothuria mammata, 68–69, 68f, 127–128
Holothuria nobilis, 45, 78, 78f
Holothuria parva, 109
Holothuria poli, 45, 67–68, 67f
Holothuria sanctori, 64–65, 64f
Holothuria scabra, 37–39, 37f, 76–77, 76f, 106, 107f

Holothuria tubulosa, 63–64, 63f, 127–128
Holothurin A, 88–89, 89f
Holothurin B, 88–89, 89f
Holothuroids, 19–21, 26
Hot air drying, 159–160

I
Immunity function, 9
Instant sea cucumbers, 162–163, 162f–163f
Intestinal segment, 7, 24–25
Intracellular digestion, 26
Isostichopus badionotus, 24–25, 25f

J
Juvenile development, 137–143

K
Kingman Reef, 46
Kruppel-like factor 2 (Klf2), 43, 44f

L
Larval development, 137–143
Leptosynapta crassipatina, 38–39
Linoleic acid, 111–114, 113f
Lissocarcinus orbicularis, 51–52, 51f
Locomotion function, 28

M
Mariculture
    areas for, 176, 176f–177f
    artificial induction, of spawning, 131–133
    broodstock, 127–128, 130–131
    embryo development, 137–143
    environmental factors, 143–144
    farming, 149–150
    fertilisation, 135–137
    gonadal development, 128
    grow-out methods, 144–146
    juvenile development, 137–143
    larval development, 137–143
    models for, 177–179, 178f–179f
    spawning behaviour, 133–134
Marketing, 179, 180f
Mechanical shock, 132
Mediterranean Sea, 61f
    Bohadschia argus, 69, 70f
    habitat preference, 69, 71t
    Holothuria arenicola, 69, 69f
    Holothuria arguinensis, 65–66, 65f
    Holothuria froskali, 69, 70f
    Holothuria mammata, 68–69, 68f
    Holothuria poli, 67–68, 67f

Holothuria sanctori, 64–65, 64f
Holothuria tubulosa, 63–64, 63f
Stichopus regalis, 66, 66f
Meristic acid, 111–114, 113f
Microwave freeze-drying (MFD), 160
Middle East, 175
Molpadiida, 1
Molpadonia, 173
Morula cells, 19
Mouth, 4

N
Nervous system, 29, 29f–30f
Nutritional values, 87–89, 89f, 111–114

O
Ocean acidification, 49
Ohshimella ehrenbergii, 108
Oleic acid, 111–114, 112f
Organic matter decomposition, 48
Ossicles, 19–21, 21f, 63
Oven drying, 159

P
Palmitic acid, 111–114, 111f
Papillae, 3
Parastichopus californicus, 50
Pearsonothuria graeffei, 4, 5f, 133–134
Periclimenes imperator, 51–52, 51f
Peripheral nerves, 29
Persian Gulf, 103–104, 103f, 104t
  aquaculture development, 110–111
  habitat preference, 109, 110t
  Holothuria arenicola, 109
  Holothuria bacilla, 109
  Holothuria hilla, 107, 107f
  Holothuria impatiens, 108, 108f
  Holothuria leucospilota, 106, 106f, 128,
    128t–129t
  Holothuria notabilis, 109
  Holothuria parva, 109
  Holothuria scabra, 106, 107f
  Ohshimella ehrenbergii, 108
  Stichopus herrmanni, 104–105, 105f, 105t
  Stichopus monotuberculatus, 108
  utilisation, 111–118
Pesticides, 174–175
Pharynx, 21–25
Polian vesicles, 28
Polyethene sheets, 139, 141f
Polynoid worms, 51–52, 52f

Polysaccharides, 9
Polyunsaturated fatty acids (PUFA), 87
Population genetics, 43–47, 44f, 46f
Predators, 53, 53f–54f
Processing methods, 157f
  canned sea cucumbers, 164, 164f
  drying, 158f
    dry-salting method, 162
    electrohydrodynamic (EHD), 161
    exportation, 157–159, 158f
    freeze, 160–161
    hot air, 159–160
    microwave, 160
    oven, 159
    solar, 159
  instant sea cucumbers, 162–163,
    162f–163f

R
Radial nerve cord, 29
Ready-to-eat sea cucumbers, 162–163,
  162f–163f
Red Sea, 61f–62f, 62, 75f
  Actinopyga mauritiana, 79–80, 79f
  Bohadschia marmorata, 82, 83f
  Bohadschia vitiensis, 82, 83f
  habitat preference, 72t–75t, 82
  Holothuria atra, 80–81, 80f
  Holothuria edulis, 81–82, 82f
  Holothuria fuscogilva, 77, 77f
  Holothuria leucospilota, 81, 81f
  Holothuria nobilis, 78, 78f
  Holothuria scabra, 76–77, 76f
  Stichopus herrmanni, 78, 79f
Regeneration
  anatomic features of, 41, 42f
  Apostichopus japonicus, 41, 41f
  digestive tube, 39, 40f
  genes, 41
  internal morphology of, 39, 39f
  intestines, 40
Rehydration, 164
Reproductive system, 33f
  asexual production, 31–32
  binary fission, 31–32, 32f
  Cladolabes schmeltzii, 32–34, 32f
  Cucumaria frondosa, 29–31, 30f
  spawning, 29–31, 31f
Respiratory system
  histology, 27, 28f
  morphology, 26–27, 27f

S

Saturated fatty acid (SFA), 111–114
Sclerodactyla briareus, 38–39
Seabed, 8–9, 8f, 10f, 10t–13t
Sea pens, 145
Secretory cells, 26
Sexual reproduction, 84
Shrimp, 51–52, 52f
Solar drying, 159
Spawning, 6–7, 31f, 127t
    Actinopyga agassizii, 6–7, 7f
    artificial induction
        algae path, 132–133
        drying, 132
        gonadal stimulation, 132
        mechanical shock, 132
        thermal shock, 131–132
        water pressure plus temperature,
          132
    behaviour, 133–134, 133f–135f
    bioturbation, 47, 47f
    broodstock, 130
    Holothuria arenicola, 84
    Holothuria leucospilota, 81
    Holothuria poli, 67
Stearic acid, 111–114, 113f
Stichopus chloronotus, 1, 2f

Stichopus herrmanni, 1, 2f, 19–21, 22f–23f, 78,
    79f, 104–105, 105f, 105t
Stichopus horrens, 127–128, 135–137, 136f
Stichopus monotuberculatus, 108
Stichopus regalis, 66, 66f
Suspended culture, 145–146
Synapta maculata, 4, 6f

T

Tentacles, 1, 19–21, 173
    Apostichopus californicus, 4, 5f
    Cucumaria miniata, 4, 5f
    Pearsonothuria graeffei, 4, 5f
    Synapta maculata, 4, 6f
    water vascular system, 28
Thermal shock, 131–132
T-shaped cells, 26
Tube feet, 3, 5f

U

Unsaturated fatty acids, 9

W

Water chemistry, 49
Water pressure plus temperature, 132
Water quality parameters, 137–138, 138t
Water vascular system, 28